In the Grip of the Distant Universe
The Science of Inertia

In the Grip of the
Distant Universe
The Science of Inertia

Peter Graneau
Formerly of Northeastern University, Boston, USA

Neal Graneau
University of Oxford, UK

World Scientific

NEW JERSEY · LONDON · SINGAPORE · BEIJING · SHANGHAI · HONG KONG · TAIPEI · CHENNAI

Published by

World Scientific Publishing Co. Pte. Ltd.

5 Toh Tuck Link, Singapore 596224

USA office: 27 Warren Street, Suite 401-402, Hackensack, NJ 07601

UK office: 57 Shelton Street, Covent Garden, London WC2H 9HE

British Library Cataloguing-in-Publication Data
A catalogue record for this book is available from the British Library.

IN THE GRIP OF THE DISTANT UNIVERSE
The Science of Inertia

ISBN 981-256-754-2

Printed in Singapore by B & JO Enterprise

To Brigitte Graneau,
whose constant love, support and encouragement
have proved vital to bringing
this book to fruition.

All truth passes through three stages:
    First, it is ridiculed
    Second, it is violently opposed; and
    Third, it is accepted as self-evident

Arthur Schopenhauer (1788-1860)

Preface

The word *inertia* is known to all, but its various definitions hide its overwhelming importance both in our daily lives and in the evolution of our universe. It is sometimes described as a reluctance to change but this is applied as often to the motion of lifeless objects as it is to the actions of human beings. It has almost become synonymous with apathy and procrastination. Yet at a deeper level, we are all subconsciously aware that the controlling hand of inertia holds a very powerful role in our mechanical world.

Historically, inertia was first regarded as a property of matter, but it became philosophically unacceptable that an inanimate object could guide its own destiny. From the early 18th century until today, this difficulty has been avoided by speaking of the force of inertia. This rarely discussed force is anything but apathetic and acts on every object which is speeding up, slowing down or changing direction.

The inertia force is so fundamental that one of the earliest experiments performed on the surface of the moon by Apollo astronauts was the simultaneous dropping of a hammer and a feather in the airless environment. The two objects of different mass struck the lunar surface simultaneously showing that it is more than the force of gravity which determines the acceleration of freely falling bodies. The force of inertia not only controls the speed of falling objects but of everything which is accelerating relative to the fixed stars. It breaks a glass when it hits the floor and pins us to the back of the seat in an accelerating sports car. It is also responsible for all centrifugal forces. In this role it prevents the moon from crashing down to earth as well as causing a disc to explode if it spins too fast.

Even though we all know intuitively how these forces feel and act, the modern physics paradigm, dominated by Einstein's relativity theories, makes it impossible to locate the matter responsible for inertia. Textbooks confront the dilemma by calling inertia a "pseudo" or "fictitious" force. According to most contemporary physicists, it is the force that isn't there!

Fortunately, there have been more reasonable theories developed in the last 300 years of post-Newtonian science which have not shirked the issue of discovering the cause of the force of inertia. The most successful of these concepts is often referred to as Mach's principle in which all of the matter in the universe becomes the cause of the force of inertia acting on every piece of accelerating matter. The creator of this theory, Ernst Mach, a distinguished Austrian physicist was one of Einstein's early heroes. However when Einstein realized that his relativity theories could not encompass Mach's principle, the two became adversaries. Since then, many physicists have been trying to brush the force of inertia under the carpet. The fact that it remains in textbooks because it is required to solve real engineering problems demonstrates clearly that there are still unresolved conflicts in modern physics.

The force of inertia is a signpost to new knowledge that underlies the laws of nature. It demonstrates the holistic aspect of the universe which is due to instantaneous action at a distance forces. Einstein objected to such a system as "spooky" and therefore unbelievable. However his conjecture of 4-dimensional curved space time is no less miraculous. We must concede that anything is possible as far as the fabric of the cosmos is concerned. What we should not condone is the academic repression of a real force just because it does not fit into a popular theory.

This book describes many of the triumphs and tribulations encountered on the path toward an understanding of this much maligned king of forces, inertia. Without hiding behind the veil of mathematics, we have endeavoured to reveal the flavour of an active controversy in which the reader can form his own opinions based on a wide body of physical evidence and experiments no longer reported in modern textbooks. Some of the foundations of modern physics are shown to be less secure than commonly depicted. The reward for seeking a deeper appreciation of the force of inertia is to discover that we are not just observers but are active contributors to the instantaneous fate of every atom in the universe.

We would like to thank Julie Hammond for her very careful and thoughtful perusal of the manuscript and Brigitte Graneau and Mayela Zamora for their patience, love and support.

Concord, Massachusetts, USA   Oxford, England

December 2005

Contents

Chapter 1

All Matter Instantaneously Senses All Other Matter in the Universe

It is easy to live on our warm and fertile planet and feel that our lives are affected only by the natural and man-made systems that exist in our local environment and that we can see with the naked eye. We know that the sun provides most of our thermal energy and that the forces of nature act to ensure that the earth maintains the right separation to support life. In a lesser way, we are also aware that our moon directly causes the ocean tides and also determines biological growth and fertility cycles. It seems however that none of the other objects in our solar system affect anything on our planet in any discernible way.

The spectacle of the night sky, with our planetary companions and the much more distant stars of our Milky Way as well as other galaxies extending to the limits of our best telescopes has certainly inspired awe in all human societies. For practical purposes, this display has been monitored closely for centuries and since the planets are unrestricted by friction forces, they have helped us understand some of the basic laws of physics. Now, astronomers use the latest data in an attempt to discover the past and future history of the universe. Is it possible however that the universe is not simply something that we observe by telescope? Perhaps we are intimately affected by our universe and because of its vast extent, it may well be responsible for very real forces that act on us, the earth and all of the objects around us.

Ernst Mach (1838-1916) [1.1], a highly regarded figure of the European scientific establishment at the end of the 19th century, believed that the force that prevents the earth from falling into the sun or that

1

squeezes us against the door of a cornering car is caused directly by every piece of matter in the universe. In fact, these forces act on every body or thing that changes its speed or direction of motion, and they are broadly attributed to a concept known as inertia.

We understand these forces instinctively when we consider that it will take more strength to throw a heavy rock than it will a lighter one. Similarly, it also requires more muscle power to slow down and catch the more massive object. These forces therefore are caused both by a change in velocity, usually called acceleration or deceleration, but also the strength of the force depends on the mass of the object. Figure1.1 shows the force of inertia operating on many different scales from the pulling apart of celestial bodies to the breaking of a plate when dropped on a hard floor.

Inertial force opposing gravity and stabilising earth's orbit

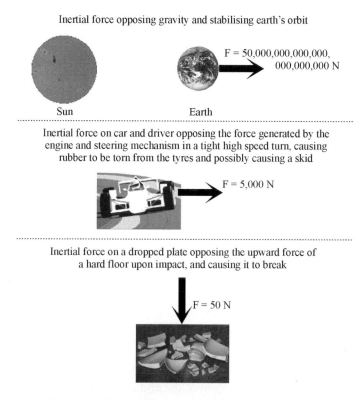

F = 50,000,000,000,000, 000,000,000 N

Sun   Earth

Inertial force on car and driver opposing the force generated by the engine and steering mechanism in a tight high speed turn, causing rubber to be torn from the tyres and possibly causing a skid

F = 5,000 N

Inertial force on a dropped plate opposing the upward force of a hard floor upon impact, and causing it to break

F = 50 N

Figure 1.1 : Three different examples of the force of inertia

The most remarkable feature of these forces is that unlike electric, magnetic or gravitational forces, we do not seem to be able to attribute their cause to any one or group of nearby objects. Even more remarkably, the force seems to appear in the same magnitude regardless of the direction of motion or acceleration. It does not matter whether a car is travelling north, south, east or west or even uphill or downhill, but if you turn to the left, then you will feel an inertial force pushing you to the right. The heavier you are, the harder your ribs will be compressed against the door.

The hypothesis of an instantaneous connection between the distant universe and the inertial forces on every object we observe is generally referred to as Mach's principle and has fascinated thinkers, scientists, and philosophers throughout the twentieth century. The principle is reminiscent of Newton's gravitational particle interactions which reach to the furthest corners of the cosmos. The earth, the moon, the planets and the sun are the dominant objects causing gravitational forces on the surface of the earth. However it must be a more orderly distribution of matter which underlies Mach's principle. Isotropic inertia forces, therefore, must rely on a remote matter distribution which due to its symmetry produces no noticeable gravitational effects on earth.

At the dawn of the scientific revolution, Johannes Kepler (1571-1630) in southern Germany and Austria and Galileo Galilei (1564-1642) in northern Italy laid the foundations of a new science called inertia. Both pioneers had royal patronage. Kepler made his most important discoveries while at the Habsburg court of the Emperor of the Holy Roman Empire in Prague, while Galileo served the Medici Grand Duke of Tuscany in Florence, the glittering citadel of the renaissance. Both were astronomers and physicists fascinated with the solar system of Copernicus which deprived the earth of its privileged position at the center of the universe. At the time, the Christian God was said to have created the earth as a stationary object to be a home for the human race. Though both men were devoutly religious, their views on the solar system brought them into conflict with the separate wings of the Christian establishment. Kepler was a Protestant and Galileo a Catholic.

The outspoken and extrovert Galileo exuded the kind of self-confidence which eventually led to his trial and condemnation in Rome.

More timid and of fragile health, Kepler avoided confrontation. A book he wrote to explain the Copernican world system was not published until after his early death at the age of 59.

Kepler and Galileo held each other in great regard and they had a long correspondence. When the professors of Padua chose not to observe the moons orbiting Jupiter through Galileo's telescope, for they thought it might undermine their scientific beliefs, Galileo wrote to Kepler:

> "what would you say of the learned here, who, replete with the pertinacity of the asp, have steadfastly refused to cast a glance through the telescope? What shall we make of this? Shall we laugh or shall we cry?".

Fundamentally, however, the two men disagreed on an issue which has remained the most enduring controversy in the history of physics: how do two particles of matter interact with each other when they are not in contact? The term "physics" was introduced by Aristotle nearly two-and-a-half millennia ago. In the first major pronouncement of this science, Aristotle contended that matter could not act where it was not. With this opinion he chased away the spirits, ghosts and Gods of ancient times and all their occult trappings. For many centuries scholars treated Aristotle as their inspired leader who had introduced the Age of Reason.

Unfortunately, it is often warfare and national defence that provides the incentive and resources for scientific discovery. This was no different in the middle of the 16th century when Giovanni Benedetti (1530-1590) studied the problem of the flight of cannon balls. During his research, he tied two objects of equal weight together with a thin thread and expected that they would now fall twice as fast as each on its own according to the principles of Aristotle. He found however, that this was not the case, and confirmed that all objects fall to earth at the same rate, regardless of their weight. Benedetti was never publicly recognized for his discoveries and it fell to Galileo to take the wrath and the fame for breaking the Aristotelian spell and proving that many of the Greek philosopher's claims were wrong. What is more, he demonstrated it with a series of simple experiments. Nonetheless, he held on to Aristotle's belief that matter cannot act where it is not. Kepler turned out to be a more adventurous spirit. By studying planetary motion he had come to the conclusion that the sun and earth attract each other and so do the earth and the moon. This mutual attraction across empty space was contrary to Aristotle's teaching.

When Kepler was 29 years old, William Gilbert (1540-1603) published his famous treatise on magnetism *De Magnete* [1.2] in England and argued that the earth was a spherical magnet. He demonstrated with many experiments how magnets attract and repel each other. It is hardly surprising, therefore, that Kepler attributed the mutual attraction between celestial objects to magnetism.

Galileo did not believe it and generally avoided the subject of attraction in his extensive writings. Cohen [1.3] attributed the following quotation to Galileo.

> "But among all the great men who have philosophized about this remarkable effect (the attraction between celestial masses), I am more astonished at Kepler than any other. Despite his open and acute mind, and though he had at his fingertips the motions attributed to the earth, he has nevertheless lent his ear and his assent to the moon's dominium over the waters (tides) and to occult properties, and such puerilities."

The profound issue which divided Kepler and Galileo is still not settled, 400 years later. Physics has progressed along a string of paradigm changes from Cartesian ether whirlpools to Newtonian instantaneous action at a distance, on to Faraday's magnetic lines of flux and Maxwell ether stresses, to be superseded by Einstein's flight of energy, curved space-time and the photon-electron collisions of quantum electrodynamics. But do we yet understand how two separated magnets attract each other?

Consider figure.1.2 which shows two horseshoe magnets sticking to the vertical sides of a copper plate. The magnets are held up, against the pull of gravity, by their attraction to each other and friction on the copper plate. If we wish to explain the attraction with modern physics, we have to call upon quantum electrodynamics (QED). One of the originators of QED, Richard Feynman [1.4], claimed it explains everything except gravitation and nuclear forces. Hence it ought to cover the attraction of two magnets.

In QED, forces between particles of matter are mediated by the collision of photons with electrons and the accompanying momentum transfer. So streams of photons must leave each of the two magnets of figure 1.2, spontaneously and forever, and then pass through a copper plate, finally colliding with electrons at the surface and deep inside the opposite horseshoe magnets. A simple collision between two particles produces repulsion, therefore in order to generate attraction between the magnets, the photons must navigate around the magnets, turn and strike them in the back.

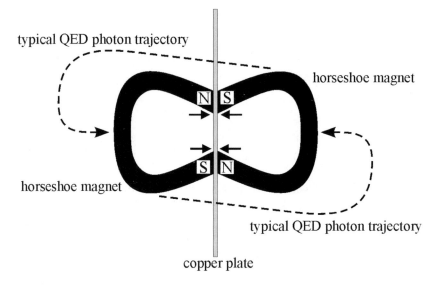

Figure 1.2 : Two horseshoe magnets attracting each other through a copper plate

This mechanism is so ludicrous that it will not be found discussed in textbooks. Nor will most professors mention it to a class of students. An exception was Guy Burniston-Brown [1.5], a reader of physics at the University of London. On one occasion he challenged an audience of students at Oxford University to explain the attraction and repulsion between two magnets without employing action at a distance. He was met with silence. Later he described this incident in his book on retarded action at a distance and wrote:

"Why should we not admit that, sometimes, what appears to be happening is happening? The refusal to accept action at a distance has led to all the difficulties and tortuous explanations connected with ether-vortices, waves, twisted space-time and many others, together with abortive experimental efforts to detect the ether. The time now has surely come to cut the Gordian Knot by abolishing all the ethers, abandoning the attribution of physical properties to "nothing", and rejecting purely mathematical constructions like space-time."

So Burniston-Brown was on the side of Kepler and did not understand the doubts expressed by Galileo with respect to attraction. Burniston-Brown made an important contribution to inertia science which will be discussed in chapter 10.

During the past century physics has become increasingly more incomprehensible. This trend has been caused by layers of esoteric mathematics which the human intellect finds difficult to translate into mechanical models and pictures. However, in order to design and usefully interpret experiments, we are forced to devise mental constructs that represent the mathematical expressions. This has led to the seemingly paradoxical properties that are accepted as an unfortunate by-product of modern physical theory. Most people feel instinctively that an electron cannot, simultaneously, be both a particle as well as an extended wave of something. Similarly, since time is an abstract concept to begin with, there is considerable confusion regarding what is meant by relativistic time dilation. Einstein's main doubt regarding quantum mechanics concerned how photons far apart from each other can instantaneously correlate their actions while adhering to the communication speed limit of the velocity of light? What makes particles increase their mass and weight with velocity relative to an arbitrary observer? Why should the orbiting planets be so exact in their wanderings while the motion of quantum particles is uncertain and erratic? Matter is supposed to be continually created and annihilated, invisibly, all around us and the whole universe sprung from a single infinitesimally small point in a Big Bang.

Whatever happened to the Age of Reason? The wonders wrought by mathematics are no less mystifying than the stories told by religions. The dependability of the miracles hinges on the answer to one of the oldest riddles of science. The British astronomer W.H. McCrea [1.6] put it succinctly. In 1971 he wrote: "If we drop an apple, it falls towards the center of the Earth, but how does it know where the center of the Earth is?" Newton would have responded without hesitation and said that every particle of the earth attracts the apple and the sum total of these attractions is directed toward the center of the earth. To achieve this, all of the atoms of the apple must sense all the atoms of the earth. They must know each other's mass and their distance of separation in order to attract each other with the correct force. Since Newtonian gravitation works so well, we Neo-Newtonians believe the mutual awareness of matter particles is not just a theory but a fact.

Einstein saw things differently. He thought the simultaneous attraction of two particles was "spooky". He would not accept that this attraction was built into matter at the time of matter creation, even though the very existence of matter is equally inexplicable. Here we have reached the very foundations of science which can only be discussed in terms of metaphysics. Einstein justified his opinion as follows [1.7]:

> "He (Newton) was also not quite comfortable about the introduction of forces operating at a distance. But the tremendous practical success of his doctrines may well have prevented him and the physicists of the eighteenth and nineteenth centuries from recognizing the fictitious character of the foundations of his system."

Einstein adhered to Aristotle's principle that matter could not act where it was not. He conjectured that matter would only respond to contact pressure from energy flying through space and striking it, or to collisions with other particles. This has become known as "Einstein local-action". According to this view the apple falls toward the center of the earth because it is running in a groove of curved space-time which presses it to follow the path also predicted by Newtonian forces of gravitation and inertia.

We can now reverse the argument and express the opinion that the practical success of field and relativity theories may have prevented the physicists of the twentieth century from recognizing the fictitious character of curved space-time. So it seems that in the end the basic question of how matter interacts with other matter apparently rests on opinions and not on experimental facts. Both opinions have been fielded by many distinguished physicists, the respective groups being led by the figureheads of Newton and Einstein.

If, however, experiments should come to light which contradict either Newton's far-action theory or Einstein's local-action theory, then one of the two world views would no longer need to be considered. Of course it is possible that both theories are flawed. Then we would have to invent an entirely new matter interaction principle, however none has come to light in the 2300 years since Aristotle coined the word "physics". We therefore seem to be landed with Newtonian mutual simultaneous far-actions, that is attraction and repulsion between separated bodies, or the collision dynamics of Einsteinian local-actions.

It has to be recognized that certain theories are not relevant to all experiments, but remain valid in the appropriate situations. For example, Newtonian mechanics does not tell us anything about the velocity of light nor about optical effects near massive objects such as the sun. This limitation of the theory represents no disproof of Newtonian gravitation. If, however, the force of inertia, which controls the acceleration of falling objects, would not act in the same way for heavy and light objects, as predicted by Newtonian mechanics, then Newtonian far-action would be flawed. Galileo had already demonstrated to all of his peers that all bodies fall toward the earth with the same acceleration regardless of their mass, and with this one fact, toppled the long-held Aristotelian view of physics. Einstein later installed curved space-time and the general theory of relativity and consequently revived the Aristotelian philosophy.

A series of recent experiments performed by ourselves and others, have been published in the scientific literature and compiled in our earlier books, *Newton versus Einstein* [1.8] and *Newtonian Electrodynamics* [1.9], which have shown that the modern Lorentz force on metallic conductors is flawed. The earliest of these demonstrations were performed by Ampère (1775-1836) [1.10], the founding father of the subject of electrodynamics. His discoveries came more than 50 years prior to the proposal of the Lorentz force which eventually succeeded Ampère's original electrodynamic force law. The newer law became popular primarily because it suited the return to the philosophy of local-action which was being actively reinstated, mainly in England in the last two decades of the 19th century. However Ampère's law has never been experimentally disproved. Since Einstein's revolution, the Lorentz force has become the only force in the modern theory of relativistic electromagnetism. Its validity is taken for granted in the recent unification of the electromagnetic with the weak nuclear force [1.11]. As well, it is thought to explain the acceleration of the metallic conductors in all of our electric machines, including the generators that satisfy our insatiable demand for electricity. The Lorentz force represents a momentum transfer by collision between an electromagnetic field and a current carrying conductor and as such is part of the Einstein local-action philosophy.

Perhaps the most important discovery that the authors have made is best discerned by the geometry of an electromagnetic device called a railgun. During the 1980's, the railgun received a lot of interest as part of the US Star Wars research program and has since been investigated as a potential artillery weapon. However, disregarding its destructive capabilities, it has provided a very revealing test bed on which to compare the competing theories of electromagnetism. The basic elements and geometry of this device are seen in figure1.3.

The metallic rails of the gun and the capacitor bank are heavy and fixed to the laboratory. The armature, however is free to slide along the rails. This is the bullet that is accelerated and eventually leaves the rails. The capacitors, which act like a very fast battery are charged so that they can be discharged very quickly through the rails/armature circuit. The force on the armature is a function of the electrical current flowing in the circuit and its momentum gain is related to the force and the time duration of the current pulse. This measurable momentum is equally well predicted by the local Lorentz law as well as the non-local Ampère force law. However, a problem surfaces when the Lorentz force has to account for local Einstein action regarding the impact of electromagnetic energy flying through the air between the rails from the capacitor to the armature. The Lorentz force must come about as a result of the collision of this flying energy (which also carries momentum) with the metal of the armature. If we believe in momentum conservation, which Newton discovered and Einstein never disputed, we know the momentum of the flying energy because it must be equal to the easily measured momentum acquired by the projectile. In Einstein's theory the energy travels with the velocity of light. It is then easy to calculate with $E=mc^2$ how much energy must have been transferred between the rails to satisfy Einstein's local-action. This mechanism turns out to require thousands of times as much energy as was originally stored in the capacitor prior to the discharge. If we also cherish the principle of energy conservation, then this is a clear violation of Einstein's local-action principle. It is a huge discrepancy which also occurs all the time in every electric motor and generator. [1.9, 12, 13]

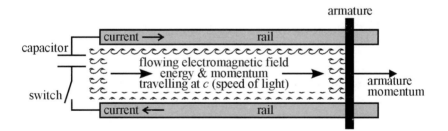

Figure 1.3 : Railgun geometry, depicting the field energy mechanism of the Lorentz force

To overlook this fact is synonymous with the blind adherence to doctrine demonstrated by the professors at Pisa in spite of Galileo's free fall demonstrations. They told their students to ignore the experiments. Aristotle had taught that heavy weights fall faster than light weights and the established academicians claimed that he could not possibly be wrong.

We claim that with the railgun experiment, in which Einstein's famous equation ($E=mc^2$) fails, we have proved the veracity of the far-action philosophy embodied in Ampère's force law which simply predicts a repulsion between the rails and the armature. By insisting on the conservation of both momentum and energy, the local-action mechanism of relativity and field theories has been disproved, at least with respect to the electrodynamic forces between metallic conductors.

The indoctrination of physics students, their blind faith in what they have been taught by their elders, and the career punishment of those who challenge the consensus metered out in textbooks, has been widely discussed and has been reported since Galileo's time. There is no need to dwell on this social phenomenon. But it should be understood that Nature has spoken, and as far as Her remarks go, She supports the Newtonian world view.

What are the implications of this view? With respect to Newton's law of gravitation, the attraction between two particles depends on both masses, their distance of separation, and the direction of the straight line connecting them. Therefore, whatever determines the strength and nature of the attraction must from the outset have information about the masses and separation of the two objects. If no outside agency controls the strength of the attraction, then each particle must have knowledge of the other's existence and whereabouts. In other words, all particles of matter must be aware of each other. They must sense each other at a distance simultaneously at the same instant of time. This is the rationale behind the assertion that all matter feels all of the other matter in the cosmos.

Einstein disagreed vehemently with mainstream physicists about the probabilistic interpretation of quantum mechanics. For thirty years he stood almost alone in maintaining that the then new quantum theory of atomic and subatomic particles must be incomplete. He reasoned that, when the missing parts would be found sometime in the future, the theory would become as deterministic as Newtonian mechanics and relativity theories. Toward the end of his life, Einstein seems to have realized that the nature of the theory of quantum mechanics stems from the principle that matter can indeed act where it is not, which is in strong disagreement with the local Aristotelian philosophy. Particles are simply aware of each other and there is no need for signaling between them, as relativistic field theory requires.

Many attempts, including Einstein's own efforts, have been unable to combine general relativity with quantum theory. Even though the subject is still being actively pursued, the dilemma persists. Quantum particles apparently interact non-locally at a distance. To have come to this conclusion must have been a great disappointment for Einstein because it threatened to destroy - to use his own words - "the castle in the air" which he had erected to defend local-action. He confessed this pessimistic outlook only to his closest and oldest friends from the Bern period during which special relativity was born. Maurice Solovine was one of the members of the Olympia Academy, a circle of friends around Einstein who met frequently to discuss scientific topics. When Solovine sent greetings to his friend on Einstein's seventieth birthday on March 14, 1949, and congratulated him on the great success of his life, Einstein replied two weeks later that he thought none of his theories would stand firm [1.7].

His best friend from the Bern years was Michele Besso who actually drew Einstein's attention to the writings of Ernst Mach and Mach's understanding of Galilean relativity. There remain 110 letters which Einstein wrote to Besso. The very last letter was dated August 10, 1954, eight months before Einstein's death. In the last sentences of this letter one reads:

> "I concede, however, that it is quite possible that physics cannot be founded on the concept of the field (local-action) - that is to say, on continuous elements. But then out of my castle in the air - including the theory of gravitation, but also most of current physics - there would remain almost nothing."

One must not confuse human opinion, however well founded on logic and mathematics, with laws of nature. There is no evidence that these laws have changed during the existence of the universe. Whoever or whatever was responsible for their formation did his work billions of years ago, long before the human race arose on earth. It seems unlikely, therefore, that the laws of nature were written to conform with anything that human brains would create as, for example, logic or mathematics.

One of Newton's opinions, which he may have held only for a short time, is often invoked against his own concept of mutual simultaneous attraction between particles of matter. No letter of science has been more frequently quoted than the one in which Isaac Newton, then a Fellow of Trinity College and Professor of Mathematics at the University of Cambridge, wrote to the young clergyman Richard Bentley on January 17, 1693. It contained arguments on proof of Deity which Bentley was to use in his Boyle Lectures intended to combat the atheism widely professed in taverns and coffee houses. At the urging of Newton, Bentley would later be appointed to the Mastership of Trinity College. In this capacity he persuaded Roger Cotes, another Fellow of Trinity College, to write the preface to the second edition of Newton's Principia [1.14,Vol.1]. This preface turned out to be a most outspoken defense of action at a distance and it had Newton's approval. However twenty years earlier Newton had written in his private letter to Bentley [1.14,Vol.2]:

"It is inconceivable, that innate brute matter, should, without the mediation of anything else, which is not material, operate upon and affect other matter without mutual contact, as it must be, if gravitation, in the sense of Epicurus, be essential and inherent in it. And this is one reason why I desire you would not ascribe innate gravity to me. That gravity should be innate, and essential to matter, so that one body may act upon another at a distance through the vacuum, without the mediation of anything else, by and through which their action and force may be conveyed from one to another, is to me so great an absurdity, that I believe no man, who has in philosophical matters a competent faculty and thinking, can ever fall into it."

Strong words indeed, but not strong facts. In this letter Newton maintained that anyone believing in instantaneous action at a distance must be mentally impaired.

Six years before the Bentley letter the same Isaac Newton published the Principia which has become the most useful scientific treatise ever written. Newtonian gravitation, for the first time outlined in this treatise, was utterly dependent on the mutual simultaneous attraction of particles. Nothing was said in the Principia which raised doubt as to the existence of action at a distance.

Ten years after the Bentley letter Newton added his famous "General Scholium" to the second edition of the Principia. In it he confessed that he had not discovered the cause of gravity "from the phenomena", and he would suggest no hypothesis which could explain this cause. Roger Cotes' preface to the second edition of the Principia strongly argued in favor of simultaneous far-actions and Newton agreed to its publication.

This story illustrates that human beings, including Isaac Newton, hold opinions which change with time. These opinions must not be confused with objective scientific facts. However creative the human brain may be, it is not a generator of unshakable experimental evidence.

A reminder of the Bentley letter will be found in the letter which Einstein wrote to his old friend Maurice Solovine in 1949 [1.7]. Just after his seventieth birthday Einstein wrote:

"You imagine that I look back on my life's work with calm satisfaction. But from nearby it looks quite different. There is not a single concept of which I am convinced that it will stand firm, and I feel uncertain whether I am in general on the right track."

Again it was a private opinion which found no expression in Einstein's scientific writings. We can only surmise that, like the Bentley letter, it concerned instantaneous action at a distance. All of his relativity theories rested on Einstein's early opinion that action at a distance was "spooky". Just like Newton, he may have thought that only insane minds could believe it. Nevertheless by the time he reached his seventieth birthday, it had become clear to Einstein that quantum mechanics required instant remote interactions. In the following decades many more physicists came to the same conclusion. The rest continue to fight a losing battle in which they have recently introduced the new term quantum entanglement as an alternative to a discussion of the now taboo phrase 'action-at-a-distance'.

The remarkable difference between Newton and Einstein was that, but for some notable exceptions, Newton adhered to his self declared creed, *hypothesis non fingo,* (I do not make hypotheses) and in public he only described experimental facts and observations. Einstein, on the other hand, was proud and indeed became famous for his boundless imagination and thought experiments. The human element in relativistic field theory has become its undoing. Newton struggled with the same questions, but he recognized that the truth of nature can only be found in the objective world outside the human brain. The lesson to be learned is that if one cannot directly observe a mechanism for a physical action, it is best not to conjecture.

A modern theoretical astronomer who deserves much admiration is Tom van Flandern [1.15], who clearly recognized that many astronomical observations are not compatible with the propagation of gravity limited to the velocity of light. He updated the calculations of Pierre-Simon de LaPlace (1749-1827) [1.16,X,vii] who was the first to demonstrate that gravitational interactions had to be at least seven million times faster than what we now call the speed of light. Van Flandern used modern data to show that in fact if gravitational interactions involved messages sent between interacting bodies, then the messages must be travelling at least at twenty billion times the speed of light in order to retain the stability that we observe in our solar system. Einsteinian relativists consider the speed of light to be the cosmic speed limit and as a result find this result very difficult to assimilate in their theory. It seems a small mental step from Van Flandern's very large gravity velocity to acceptance of simultaneous remote particle interactions. LaPlace made this leap but Van Flandern is not prepared to take this step because of the human element and what he calls logic. Instead he postulates, by hypothesis, the existence of gravitons which travel much faster than photons.

Van Flandern reasons Newtonian attraction is illogical because one particle must be the cause of gravity and the other particle must feel the effect a little later. This kind of causality is clearly a human hypothesis and not a demonstrable fact of the objective world. Here we come face to face with the question of whether the maker of the laws of nature was in awe of man's intellectual power and logic? Van Flandern's logic, Newton's remark about insane minds, and Einstein's spookiness are not cold objective experimental facts which compel us to believe that two particles of matter cannot attract or repel each other at a distance. We need something better to establish the interaction principle by which nature abides.

A most remarkable example of the interference of the human brain with objective nature is provided by the concepts of space and time. When asked out of the blue, almost every adult human being will say that he or she knows what is meant by space and time. Yet there exists not a scrap of objective empirical evidence that either entity exists at all. Faced with the experimental situation, the honest scientist should admit that space is nothingness and so is time. Experimental observations deal only with the relative measures of space and time. They are the distance between material objects and the intervals between material events. We know how many measuring sticks can be laid end to end between two stones. Thus distance becomes a ratio between something variable and something that we are familiar with and that we believe remains constant such as the standard metre stick stored safely in a glass case in Paris. This ratio is just a number based on observations of objects and is certainly not something that can be called space. Similarly, our measurement of time always represents the ratio of the period of a cylical event that we are familiar with and another event. For instance, we observe that the sun rises 365.25 times in the interval that it takes for the earth to go once around the sun. This is a ratio of intervals and not time itself, but is enough to give us the feeling that we know how long a year is.

Immanuel Kant (1724-1804) [1.17], the German philosopher addressed these problems. Fifty years after Newton's death he said-in English translation-:

" .... Therefore, we shall understand by *a priori* knowledge, not knowledge independent of this or that experience, but absolutely independent of all experience."

Then he goes on to explain that space and time are *a priori* knowledge stored in our brain before birth, while what we know of the objective world is observed by the senses and then stored in the brain *a posteriori*. Why should nature find it necessary to store the *a priori* notions of space and time in our brains? A likely explanation is that without them we would not be able to sense motion and could not interpret our flexible bodies and our ever changing environment.

If we are to remain true to Newton's dictum and only create mathematical equations which model the features of the universe that we can observe, then we have to base our physics on the only motion that we can perceive, which is relative acceleration between a pair of objects. Any two objects certainly exert a gravitational force and possibly an electrostatic or electromagnetic force on each other and if they are free to move, they consequently accelerate toward or away from each other. By observation, Newton discovered that there are attractions for which the relative acceleration is related only to the masses of the two objects. He described these motions and related them to a force which he called gravitation. Coulomb and Ampère observed that charge and electric current also affect the relative acceleration of objects and ascribed their relative accelerations to electrostatic and electromagnetic forces respectively. These three pioneering scientists were all able to empirically discover non-local force laws which describe observed accelerations without any consideration of the mechanism by which they acted.

Another type of force is however also detectable. This is the force of inertia. It can be generalised as a force which counteracts any acceleration of an object with respect to the frame that Mach described, in which the bodies of the distant universe are observed to be at rest. This instantaneous force appears to be related to the mass of the object and its acceleration with respect to the Machian frame. An interaction between an observed object on earth and one in the distant universe must be a non-local interaction. The stars in our galaxy are far enough away from us that however fast they are moving, they form a virtually fixed background upon which we measure the motion of our much closer planetary companions. In the same way the relative motion between the galaxies other than our own also can be considered a fixed background relative to which we can measure acceleration. Further, since the laws of inertia appear to be the same for all directions of motion, then we can assume that the parts of the universe that significantly contribute to the inertia force are distributed uniformly in all directions. This is called an isotropic distribution.

The most familiar manifestation of the force of inertia is the linear resistance to acceleration. This is the force that appears whenever an object is subjected to an applied force either by contact or by gravity, electrostatics or electromagnetism. The inertia force precisely opposes the applied force in such a manner as to allow a finite and predictable acceleration. It is the reason that all objects fall toward the earth with the same acceleration regardless of their mass. If this force did not exist, then any applied force would produce an infinite acceleration and the universe would have collapsed long ago due to the force of gravity.

If however, a force is applied to an object which is already moving perpendicular to the direction of the applied force, then the inertial opposition becomes known as a centrifugal force. This is the force that stretches and sometimes breaks a string used to swing a weight around our head. It is also the force that pushes a race car off a high speed corner and most importantly prevents the earth from falling into the sun.

Possibly due to the fact that the inertia forces are so uniform and also that a search for their source implies the currently unfashionable non-local interaction principle, they have been treated differently from the other forces in modern physics textbooks and are often only described as "pseudo-forces"[1.18]. Part of the problem with the image of inertial forces is that nobody has yet proposed a Newtonian non-local force law which can give the inertial force the same "true-force" status as the laws of gravitation, electrostatics and electrodynamics. Such a law is proposed in Chapter 12 of this book. Like its predecessors, the laws of Newton, Coulomb and Ampère, it makes no assumptions regarding the mechanism of non-local interaction, but simply aims to fit the observed acceleration measurements.

If indeed, objects are directly pushed and pulled by all of the bodies in the universe, then a perfect demonstration of these forces is the space compass, better known as a gyroscope as shown in figure1.4. Once the axis of a flywheel is aligned to point from one fixed galaxy to another and is held in gimballed bearings which are secured to a space capsule, it will point in this direction forever, whatever maneuvers the space ship will perform, so long as the gyroscope is kept rotating and no electric, magnetic, or contact forces can apply a torque to the axis of rotation. The atoms simply feel where the fixed stars are and are pushed and pulled by them. It is not the inertial or gravitational interaction with the nearby stars that stabilizes the gyroscope alignment. It has to be other isotropically distributed matter arranged in an unchanging way with respect to our galaxy. Every time we become aware that we are accelerating, it is because the distant universe noticed it and has pushed us.

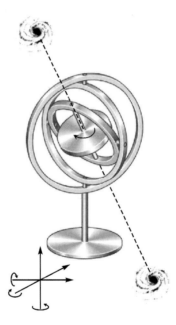

Figure 1.4 : A rotating gyroscope in gimballed bearings. The axis of rotation remains aligned with two galaxies, however the base is moved.

Chapter 1 References

[1.1] E. Mach, *The science of mechanics*, 6th ed. La Salle, IL: Open Court, 1960.

[1.2] W. Gilbert, *De Magnete*. New York: Dover, 1958.

[1.3] I. B. Cohen, *The birth of a new physics*. New York: Norton, 1985.

[1.4] R. Feynman, *QED*. Princeton: Princeton University Press, 1985.

[1.5] G. Burniston-Brown, *Retarded action at a distance*. Luton: Cortney, 1982.

[1.6] W. H. McCrea, "Doubt's about Mach's principle," *Nature*, vol. 230, p. 95, 1971.

[1.7] A. P. French, (Ed.) *Einstein: a centenary volume*. Cambridge, MA: Harvard University Press, 1979.

[1.8] P. Graneau, N. Graneau, *Newton versus Einstein*. New York: Carlton Press, 1993.

[1.9] P. Graneau, N. Graneau, *Newtonian electrodynamics*. Singapore: World Scientific, 1996.

[1.10] A. M. Ampère, Théorie mathématique des phénomènes électrodynamiques uniquement déduite de l'expérience. Paris: Blanchard, 1958.

[1.11] S. Weinberg, "Conceptual foundation of the unified theory of weak and electromagnetic interactions," in *Nobel lectures in physics 1971-1980*. Singapore: World Scientific, 1992.

[1.12] P. Graneau, "Nonlocal action in the induction motor," *Foundations of physics Letters,*, vol. 4, p. 499, 1991.

[1.13] N. Graneau, "Have you seen the light ?," in *Instantaneous action at a distance in modern physics: "pro" and "contra"*, A. E. Chubykalo, V. Pope, and R. Smirnov-Rueda, Eds. Commack, NY: Nova Science, 1999.

[1.14] I. Newton, *Principia*, F. Cajori (Ed.). Berkeley: University of California Press, 1962.

[1.15] T. van Flandern, *Dark matter, missing planets and new comets*. Berkeley, CA: North Atlantic Books, 1993.

[1.16] P. S. LaPlace, *Mècanique Cèleste*, vol. IV. Boston: Little Brown, 1839.

[1.17] I. Kant, *Critique of pure reason*. London: Macmillan, 1963.

[1.18] A. P. French, M. G. Ebison, *Introduction to Classical Mechanics*. London: Chapman & Hall, 1986.

Chapter 2

Johannes Kepler

The Astronomer who Coined the Word Inertia

What is inertia? This question was frequently asked at the dawn of the scientific revolution in the late 16th century. The answer remains elusive to this day.

If it were possible to flick a switch and turn off inertia, the universe would collapse in an instant to a clump of matter. Assuming, as we do, that the universe contains as many positive electric charges as it does negative charges, there would exist no electric forces in the matter clump to spread it out again. Similarly, nature provides an identical number of north and south magnetic poles and thus magnetic forces would also be unable to prevent the big crunch.

Fortunately, our universe also contains heat, which is the result of the motion of particles. This motion represents energy which is called kinetic energy. It is a man-made accounting strategy which represents the inertia of moving particles. Without inertia, kinetic energy and heat would cease to exist. Everything would cool down to the lowest possible temperature of the universe known as absolute zero temperature. All motion would cease and the arrangement of matter would never change. Consequently there would be no way of measuring time and life as we know it would be absolutely impossible. Is there anything in our world that is more important to us than inertia?

The German astronomer Johannes Kepler (1571-1630) was the first to use the term *inertia*. He derived it from the adjective *inert* which is of Latin origin. Its common meaning is idleness or a certain resistance to change in the motion of bodies and particles. Unfortunately Kepler did not clearly distinguish between motion and acceleration and deceleration. which has remained a source of confusion when trying to understand his work.

More than half a century before Newton, Kepler had a fair understanding of the attractive forces between celestial bodies as well as what is meant by mass and weight. While still a young student in Tübingen, Germany, with his teacher Michael Maestlin, he recognized, without the help of a telescope, that the moon consisted mainly of the same substances as the earth. He studied the lunar landscape as best he could with the naked eye and spotted the existence of lunar mountains. His professor taught him how to determine the approximate height of the mountains from the shadows they were casting.

Figure 2.1 : Johannes Kepler

Years later, after Galileo had made himself a telescope and discovered the moons of Jupiter, Kepler informed the Italian astronomer of his knowledge of the geography of our moon. Kepler speculated about space travel and wrote to Galileo:

> "So, for those who will come shortly to attempt this journey, let us establish the astronomy: Galileo, you for Jupiter, I for the Moon."

One of Kepler's biographers, Caspar [2.1], relates that the young theology student in Tübingen was stimulated by the hypothesis of the Polish priest and astronomer Nicolaus Copernicus (1473-1543) who had died fifty years earlier. This paradigm shifting and highly controversial conjecture proposed that the sun was the center of the universe, and not the earth. It stood in stark contradiction to the Ptolemaic system taught to Copernicus some 1400 years after it had been proposed by the distinguished Greek astronomer Claudius Ptolemy. The Christian Church had adopted the Ptolemaic philosophy in order to substantiate the view that God had created the earth for the benefit of man. Therefore the Copernican astronomy, which made the earth far less important than the sun, was not to the liking of the theologians of Tübingen. But Kepler's mentor Maestlin, being an enquiring observer of the starry sky, introduced his student to the Copernican system and unwittingly reared a pioneering rebel of science. There is no way of charting progress in the search for truth in nature other than by questioning and overcoming ingrained beliefs and prejudices. All great scientists have been unpopular revolutionaries, some of them for longer than others. Einstein managed to convince his peers in a decade or two, but Copernicus had already been dead for ninety years when Galileo lost his battle with the Pope over the hypothesis of the sun-centred world. In all it took about 150 years before the Copernican based astronomy became acceptable and could be openly taught to students.

It was a custom in Kepler's Lutheran Seminary in Tübingen to hold debates, or disputations as they were called, between students in front of their professors. Kepler led a disputation in which he suggested that the apparent motion of the fixed stars around the earth was actually due to the spin of the earth about a north-south axis. With the fixed stars as a background, he also discussed the Copernican orbiting of the earth around the sun. On the issue of the center of the universe, the Protestants held onto the Ptolemaic beliefs as firmly as the Catholics. Lear [2.2], another biographer of Kepler, believes that the young rebel was not given a pulpit of the Lutheran Church at the end of his theological studies because of his preference for the sun-centred astronomy of Copernicus. At the age of 23, before he sat his final examinations, he was actively encouraged to move to Graz, to take up the post of Mathematicus of Styria, an Austrian province. Although he felt ill equipped to teach mathematics and astronomy in a Protestant school, it gave him financial independence and allowed him to devote more of his attention to astronomy. He moved on the condition that he would be able to return and continue his study of divinity at a later date. However after a difficult first year, he began to make discoveries which soon dispelled his religious aspirations.

Throughout his life Kepler was at odds with the intellectual establishment represented at his time by scholars and leaders of the Christian Church. The church hierarchy could have easily ruined his career as a mathematician and astronomer. He did not have the means to practise science without financial support from others and therefore had to disguise his scientific convictions best he could. Already in the 1590's he started writing about a journey to the moon, describing clearly how space travelers would see the Copernican world. The Copernican picture of the sky seen by an observer in space should have been the subject of Kepler's first book however he did not dare to give these views an air of certainty and disguised them in what he said was a dream. The book, written in Latin, was finally published four years after his death under the title *Somnium* (Dream). It became well known as a major work of lunar astronomy.

Kepler feared that the allegorical account might confuse his readers. Throughout his life he added a long list of footnotes to the manuscript. The footnotes eventually occupied more pages than the main text. Lear [2.2] interpreted Kepler's dream in terms of modern astronomical concepts. For historical perspective, it is most intriguing to consider Kepler's speculation of how the trip to the moon would be accomplished.

First of all he thought that in order to protect the travelers from burning up in sunlight, which was not filtered by the atmosphere, the journey would have to take place during a lunar eclipse. Then the spaceship could travel the entire distance in the shadow of the earth. Kepler had a good idea of the distance to the moon and thought this had to be traversed in approximately four hours. Rockets were not yet invented and consequently he suggested that the space traveller would have to be launched with a cannon. Anticipating the effects of the force of inertia, he wrote:

"The first getting into motion is very hard on him, for he is twisted and turned just as if, shot from a cannon, he were sailing across mountains and seas. Therefore, he must be put to sleep beforehand, with narcotics and opiates, and he must be arranged limb by limb, so that the shock will be distributed over the individual members, lest the upper part of his body is carried away from the fundament, or his head be torn from his shoulders."

Not all of this early description is physically correct, but it certainly recognizes the all important existence of destructive forces of inertia. Kepler correctly reasoned that after the violent take-off, the gravitational pull-back by the earth would gradually decrease, as the traveler receded from our planet, until the attraction from the moon took over. He also correctly estimated that the speed achieved after such a launch would be too slow to complete the journey in four hours. To rectify this he fell back on allegory and proposed the spaceship may be speeded up by "will-power". Fully aware of this fictional aspect of the journey to the moon, which was merely an excuse to write about the appearance of celestial bodies according to Copernicus, Kepler also sent "will-power" ahead of the astronaut to catch him upon landing on the moon. This was his hopeful mechanism to cushion the fall of the astronaut and stop the force of inertia from breaking his bones.

All this was written fifty years before Newton published his laws of motion and put the forces of gravity and inertia on a mathematical foundation. It revealed Kepler's approximate understanding of the science of dynamics which was to come. The word *inertia* actually appeared in one of the footnotes of *The Dream*. There Kepler explained that will power alone could not overcome the effect of inertia and added:

> "There is need of some force also. For every body by reason of its own matter has a certain inertia in regard to motion which provides repose to the body in every place in which the body is placed in a position beyond the powers of attraction. Whoever would move this body from its place must overcome this force, or, rather, this inertia."

It is important to realise that Kepler here spoke of inertia as a force, rather than the mass of the body, which has also been called inertia by others. It surprised him as much as it has dozens of generations that followed him, that the force of inertia acting on a body did not depend on where the body was situated, nor the direction in which it was moving. This has remained the riddle of inertia which distinguishes the force of inertia from the force of gravity. The two forces together, in Newton's hands, caused the scientific revolution which allowed the development of our modern industrial world. Most scientists living at the beginning of the twenty-first century are as puzzled by the force of inertia as Kepler was. Many have come to believe, quite wrongly, that it is a fictitious force, in other words a force that does not exist, even though it can break things and holds the earth in its stable orbit around the sun!

The last few years of Kepler's life were dominated by the trauma of trying to publish and sell his books in a period dominated by the worst ravages of the Thirty year war. [2.3] As a Lutheran, he was forced to move from town to town, often travelling with printing presses and half completed manuscripts. He sold his books at markets and was never sure how to fund himself and his family from whom he was often separated. With one foot still in the medieval world, Kepler was able to garner patronage as an astrologer although he was careful to not overstate his predictions. Nevertheless, his final career appointment was Court Astrologer to the Duke of Wallenstein in his newly acquired Duchy of Sagan in Silesia. It was here that Kepler had to procure a printing press and once again attempt to finish the publication of The *Somnium*. One of the authors of this book, PG, went to school in this small market town of 20,000 inhabitants and for periods of time he boarded in a house on cobbled Keplerstrasse. On the house next door was a plaque commemorating Kepler's stay there. Unfortunately Kepler never finished his task and Wallenstein never paid his promised salary. Forced to travel to recover lost revenue, Kepler set off again, never to return to his family and died in Ratisbon in November 1630. The *Somnium* was finally published in 1634.

Inertia has two aspects. First, it underlies Newton's first law of motion which states that a body will coast along a straight line at constant velocity unless acted upon by an external force. The second and more important aspect is that when an external force accelerates the body, it is met with an equal and opposite force of inertia which controls the magnitude of the acceleration when measured relative to the fixed stars. Historically these two facets of the force of inertia led to Newton's first and second law of motion. Neither Kepler nor Newton were fully aware of the relationship of the acceleration with respect to the fixed stars. This was not discovered until much later by Ernst Mach. in the late 19th century.

With the description of his lunar voyage, Kepler revealed his qualitative understanding of the force of inertia. It acted in addition to the weight forces of gravity and actually added to the weight during lift-off from the earth. In the development of physics since Kepler, the two distinct forces of gravity and inertia have become merged in Einstein's general theory of relativity.

Kepler did not realize that an un-accelerated object would just drift along without being driven by any force. We now know that when a test body is not subject to acceleration, deceleration or a change in direction, it does not experience a force of inertia. Newton's first law of motion, usually referred to as the law of inertia, is therefore redundant. It merely states the fact which is inherent in the second law, namely that a body which is not subject to an external force does not accelerate which means that it coasts along in a straight line (relative to the fixed stars) at constant velocity. At such an early stage in the history of quantitative science, Newton simply had not grasped the importance of the relationship between observable objects and the distant fixed material universe. This quite excusable lapse caused him to spell out his first law of motion. A more complete discussion of the development of Newton's laws of motion with respect to the force of inertia can be found in Chapter 5.

Most of the ancient Greek texts, including the works of Aristotle, only became known in Europe during the 12th century. At this time, a rather irregular stream of manuscripts translated from Arabic sources started to illuminate medieval Europe and in this manner the Western World recovered its own intellectual heritage. According to Koestler [2.3], the early kindling that was to ignite the wildfire of intellectual development in Europe was the rediscovery of Aristotelian texts. St. Thomas Aquinas (1224-1274), a Dominican Friar and Theologian is known as an early pioneer of the belief that reason and knowledge can lead to the mind of God. He championed the work of Aristotle which led to the adoption by the Christian Church of the world picture developed 1600 years earlier in ancient Greece. After a further 300 years, Kepler was still under the influence of Aristotle who had claimed that motion of a body at all times requires a driving force. The German astronomer failed to notice the tendency of matter to move forever, unless influenced by another object. At the turn of the sixteenth century he was blindly groping for the new physics which Newton was soon to create. Nevertheless Kepler furnished Newton with two invaluable pieces of information. From Tycho Brahe's (1546-1601) most careful and time-consuming measurements of the orbits of the planets, specifically that of Mars, Kepler derived three mathematical relationships which could be used for the calculation of the positions of the planets at any time in the past or future. Without Kepler's three laws, as they became known, Newton might not have been able to convince himself of the validity of what was to become his universal law of gravitation. The three laws of planetary motion required a force of attraction between heavenly bodies which was very much like the force of attraction between two magnets. Kepler actually believed it was the magnetism of the earth and the moon which lifted the water of the oceans to produce tidal flows. Since it took centuries to clarify what was meant by magnetism, Kepler may be forgiven for having confused it with the forces of gravity. Whatever the cause of the attraction, it was important to the theories of both Newton and Kepler that this attraction was a simultaneous mutual interaction of two bodies or particles of matter.

The other idea which Kepler planted in the minds of his successors was that there had to exist forces of inertia which kept the attracting and orbiting celestial bodies apart and thereby prevented them from falling into each other. This particular form of the force of inertia is now called *centrifugal force*. Much more had to be accomplished by Newton with regard to the quantitative laws of the forces of gravity and inertia, but without Kepler's concepts of instantaneous attraction and inertial resistance to acceleration, it would not have been possible to erect the enduring edifice of Newtonian dynamics.

Newton is often quoted from a letter to his bitter rival Robert Hooke in Oxford University. In an effort to deny any assistance from his fellow Englishman, Newton cynically wrote, "If I have been able to see further, it was only because I stood on the shoulders of giants". Hooke had claimed that he had been the first to propose a law of universal inverse square gravitation and it appears that Newton used all of his political might in the Royal Society to discredit his rival's claims. Nevertheless, Newton was clearly aware of the debt he owed to his predecessors, especially Kepler and Galileo.

Johannes Kepler also stood on the shoulders of a giant. This colossus was the Danish astronomer, Tycho Brahe. Many Danes believe that Brahe was their most important scientist. Without Brahe scientific progress would probably have been greatly delayed. He discovered the 1572 'Nuova' (new star) in the constellation of Cassiopeia. King Frederick II of Denmark rewarded him for this with the grant of the Island of Hven near Copenhagen. On this Island Brahe set up the best astronomical observatory of his time. It enabled him to amass an enormous amount of very accurate data of the positions of fixed stars and the trails of planets and comets across the night sky. His meticulous and painstaking labor extended over a period of twenty years from 1576 to 1596. This was before the invention of the telescope. His principal instruments were staffs equipped with what amounted to gun sights. Large precisely marked metal rings were used to measure the horizontal and vertical angles of the pointer to the observed star. He used some of the most accurate clocks in the world and by making up to four simultaneous measurements of a star's position, he was able to attain unprecedented precision. The accuracy of his angle measurements was ± 1 minute of arc which represented an improvement by a factor of almost 10 over his contemporary astronomers. As well, he was the first astronomer to take into account the distortion due to the atmosphere, but most importantly he boldly attempted to map the entire observable universe.

Figure 2.2 : Tycho Brahe

Similar attention to detail was paid to the measurement of the positions and brightness of the planets. It was this voluminous archive of recorded data which the Imperial Court Mathematician, Johannes Kepler, inherited after Brahe's death while in the employ of the Austrian Emperor in Prague. Without it Kepler could not have derived his three laws of planetary motion, and Newton would have found it nearly impossible to convince anybody about the forces of gravity and inertia.

Figure 2.3 : Uraniborg, Tycho Brahe's observatory on Hven

Tycho Brahe watched the stars at night and talked endlessly during the day to a stream of visiting scholars. Students of the sky came like pilgrims to Uraniborg [2.4], the castle of astronomy which Tycho had built with Royal help on the Island of Hven. It was run like a small self sufficient country with Brahe acting often as a despotic squire. His empire included forty farms, and his institute had its own printing press and paper mill. The castle contained its own pharmacy including an alchemist's furnace as well as artificial fish ponds, a game reserve and even a prison for drunken and disobedient tenants. Despite its extraordinary social conditions, it was the forerunner of the great astronomical observatories of more recent times. It is estimated that the Danish king invested 1% of his annual expenditure on the development of Brahe's observatory. This can be compared to NASA which consumes only 0.4% of the US government budget. Modern facilities boast a vastly more sophisticated armoury of instruments, but the adventure of deciphering the cosmos is no more fascinating now than it was in the sixteenth century. Encouraged by Brahe's achievements, a number of universities added astronomy to their curriculum.

There is little doubt that Brahe was one of the most thorough observers of the heavens. He did not work so hard merely to compile reams of numbers, but was driven by the restless ambition of creating an improved map of the universe. In his time the world comprised the solar system and a spherical backdrop of fixed stars which did not move relative to each other. As each of these fixed stars was recorded, it's position was engraved on a five foot diameter polished brass sphere which served as the first three dimensional map of the heavens. As for the nearby planets, the Tychonic system, which he eventually proposed, was a compromise between the models of Ptolemy and Copernicus. Brahe could not divorce himself entirely from Aristotle's physics and what seemed to him a perfectly reasonable claim of the Christian Church that the earth was at rest at the center of the world. Church leaders were at the time preaching that the earth was created to benefit man. Why would God install his most worthy creation, the human race, on an insignificant planet wandering aimlessly through the cosmos? Those were precisely the theological grounds on which the Pope later condemned Galileo. The aristocratic and conservative Tycho Brahe was not psychologically prepared to deny such popular religious beliefs.

In the Tychonic system the moon and the sun revolved around the stationary and not spinning earth. This agreed with Ptolemy. However he could not deny that his observations clearly showed that Mercury, Venus, Mars, Jupiter, Saturn, and other celestial objects including comets, orbited around the sun. The whole of the solar system was surrounded by a hollow sphere which contained the fixed stars and spun about an axis which passed through the center of the earth and the polar star. The most remarkable feature of Tycho's astronomy was that this complex arrangement of planets revolving around the sun which, itself, revolved around the earth, was apparently in agreement with all his angular measurements of the paths of celestial objects.

In the two-year spell, from 1599 to 1601, while Kepler was Brahe's assistant in Bohemia, the Danish astronomer was wary of Kepler's great mathematical genius and only conceded restricted access to his hard won information about planetary motions. However after Brahe's death, in 1601, Kepler received what he had prayed for and was able to investigate the entire archive. Without it, the German would have remained a comparatively obscure mathematician and astrologer. Kepler was very aware of his historic mission and immensely grateful to his deceased master Tycho Brahe. He wrote in glowing terms about the Danish astronomer and fulfilled his pledge to Brahe to try and fit the observational data to the Tychonic world model. Eight years after Brahe's death, when Kepler presented his own new astronomy in book form, he had to admit to a discrepancy in the orbit of mars of eight minutes of arc between Brahe's observations and theory. One complete revolution contains 360 degrees or 21,600 minutes of arc. The angular disagreement was less than four parts in ten thousand. Nevertheless, this flaw turned out to be enough to disprove Brahe's world system, as subsequent developments would show conclusively. In the end it was Brahe's experimental perfection which faulted his own theoretical creation.

Another aspect of the Tychonic world which did not fit in with careful observations was the shape of the planetary orbits. Kepler discovered that they were not circles but ellipses. This was contrary to the perfectly circular orbits assumed by Ptolemy, Copernicus, and Brahe. Along with the elliptical orbits came the realization that the tangential velocity of the planets was continuously changing as they approached or receded from the focus of the ellipse at which the sun was located. Greek philosophy and Christian doctrine had both paid homage to the perfection of uniform circular motion. Scholars were reluctant to give this up. Kepler had to use all of his ample political skills to ward off prosecution while telling the truth about elliptic orbits of the planets.

In his philosophical investigations, Tycho Brahe had discussed an issue which revealed his views on the science of inertia. In his cosmology, the earth did not move at all. He thought this was confirmed by the following experiment. When a stone is thrown vertically upward from a marked spot on the surface of the earth, it always falls back to precisely this spot. The Aristotelian model predicted that if the earth were in motion, then a force must continuously propel it and everything that is connected to the earth. Throwing the stone into the air breaks the rigid connection with the earth and according to Aristotle, the stone will then not feel a force which makes it move with the earth. Therefore, if the earth was moving, then the surface of the planet should move sideways while the stone traveled up and down and consequently the stone should not land on the same spot from where it was catapulted upward. The fact that it did land precisely on this spot proved to Tycho Brahe that the earth was stationary.

For practical reasons, this particular experiment is difficult to perform. It is much easier to drop a weight from a tower and see where it lands. A plumb line to the ground would determine the spot on which the weight should land if, as argued by Brahe, the earth was stationary. To the limits of early 17th century measurement techniques, it was found that freely falling heavy objects always descend vertically down to the ground.

Today we know that the orbital velocity of the earth around the sun is 30 kilometres per second. The surface velocity due to the spin of the earth is much smaller, of the order of 0.4 kilometres per second. Let us take the example of the Leaning Tower of Pisa from which Galileo may have dropped cannon balls at about Brahe's time. The height of the tower is 55 metres and it takes 3.3 seconds for an object to reach the ground. In this time the earth's surface should have moved on for many kilometres from where it was at the instant the cannon ball was released. In practise the balls always fell to the foot of the tower. From the Aristotelian perspective, Tycho Brahe did seem to have a good point.

There was also another cannon ball argument. If, as Kepler had maintained, the daily motion of the sun across the sky was actually the result of the earth revolving about a north-south axis, with the surface of the earth rotating from west to east, a cannon firing identical shots in the westerly and easterly direction should cover a greater distance on the surface of the earth when aiming west. This was not found to be the case. It once more supported the views of the Danish astronomer.

As reported by Thoren [2.4], Tycho Brahe actually tested his hypothesis that movement of the earth's surface would determine where a falling object landed. Brahe wrote:

"Some people think that if a missile were thrown upward from the inside of a ship, it would fall in the same place whether or not the ship were moving. They offer these assumptions gratuitously, for things actually happen quite differently. In fact, the faster the motion of the ship, the more difference will be found."

With today's knowledge, one can only surmise that the motion of Tycho Brahe's ship was not very steady and the "difference" was caused by the rocking of the boat. He must have heard of inertia from Kepler, but gained no real understanding of the concept of inertial force. It was simply too early in the age of inertia physics, a science that is still evolving four centuries later,

The Dane failed to appreciate both aspects of inertia: the force-free coasting of a body and the inertia force which resists acceleration. The cannon ball falling from the Leaning Tower coasted along with the tower through space and, hence, always fell to the foot of the tower. With the technical deficiencies of the era this experiment was not a good test of the state of motion of our planet, as Tycho Brahe had assumed. It seems however that Kepler was also unaware of this facet of inertia. It remained to be discovered by Galileo in Italy.

It is of historical interest to note that the dropping of objects from towers has now been used to measure the motion of the earth, however it is a much smaller effect than would have been predicted by Brahe. Since the earth is rotating in the west to east direction, the top of a tall tower has a higher eastward velocity than its base. Therefore when an object is "dropped" from the top of the tower, it is also inadvertently thrown a bit horizontally relative to the base of the tower. As a result, the object always lands slightly to the east of the spot, which is found by dropping a plumb line from the top of the tower. The amount of eastward deviation depends on the height of the tower and the latitude of the experiment. For example, a ball dropped from the Leaning Tower of Pisa will deviate from the plumb line by approximately 5 mm. Experiments performed in the first few years of the 20th century confirmed this prediction and are described in [2.5]. It is ironic that an experiment that Brahe conceived to prove that the earth is stationary should, 300 years later, actually demonstrate the opposite and prove conclusively that the earth is spinning.

Despite their mistakes, in the fifty years between 1580 and 1630, Tycho Brahe and Johannes Kepler transformed astronomy from a mystical medieval subject, still intermingled with astrology, to a science of exact observation and mathematical prediction. At the same time, both continued to humour their patrons with astrological forecasts and oracular predictions in order to maintain their interest and to garner fame and financial rewards for themselves. During this period, Brahe first and then later, Kepler, laid the corner stones on which Newton was to build his mechanics of gravitation and inertia. Incomplete as their understanding was, the two colorful personalities of the late renaissance had a powerful influence on how man was to view nature in centuries to come.

Their astrological forays were in general not very successful. Gade [2.6] mentions that in 1566, the young Tycho Brahe posted the horoscope of Sultan Suleiman the Magnificent, the Ottoman Emperor, on the walls of his university. He claimed that a recent lunar eclipse had predicted the Sultan's early death. Subsequently the apprentice astrologer learned that the Sultan had already died six weeks earlier.

Kepler had a bit more luck with his attempts to foretell future events. He was asked to compute the positions of the planets at the hour of birth (nativity) of Albrecht Wallenstein, the most famous general of the thirty year war. It was Wallenstein's patronage that later brought Kepler to Sagan in Silesia in 1628. From the nativity provided by Kepler in 1624, a French professional astrologer cast Wallenstein's horoscope. It was predicted that in March 1634 there would be "dreadful disorders over the land" [2.3]. Kepler died in 1630 and Wallenstein was assassinated on February 25th 1634.

Kepler and Brahe were very different personalities. The German was frail and of impaired eyesight. He was born prematurely and contracted smallpox at the age of four, nearly going blind. His father was a soldier in the endless wars of Germany. His mother became a camp follower, leaving Johannes to the chaotic care of his grandparents. Throughout his life Kepler felt insecure and had to compromise his scientific convictions in order to earn a living. One of the more distressing episodes was the defense of his mother who was on trial, at the height of the witch hunts in Germany, in the early seventeenth century. He proved the spells that his mother was supposed to have cast on people were really arthritis, lumbago, and similar afflictions. When arriving in Weil-der-Stadt, his birthplace near Stuttgart, he found his mother chained to the walls of a cave and he was accused of heretical beliefs in astronomy. Yet in his systematic way of solving problems, he won the legal battle for his mother and spared her literally at the last minute from the machines of torture.

In stark contrast, Tycho Brahe exuded the arrogance of an aristocrat and was of robust health. In the same year as his unfortunate astrological embarrassment, the rumbustious Dane had a calamitous encounter with another young nobleman. At a Christmas party the two quarrelled about a mathematical problem. The conflict became so intense that they adjourned to the darkness of night for a sword duel. In the course of the steely confrontation, Tycho Brahe had most of his nose cut off, which swiftly settled the dispute. Plastic surgery not being what it is today, ordinary mortals would have opted for a wax nose. However, the wealthy Dane chose to have one beaten out of gold and silver. When it was painted and stuck on firmly, one could hardly notice the prosthetic fix. He is said to have always carried a box of glue in his pocket in case the nose started to shift.

His golden era in charge of the world's largest and most sophisticated observatory came to an end when at the age of fifty, he fell out of favour with the new King of Denmark, Christian IV. Brahe had left letters from the king and high courts unanswered and was costing the country too much money. He therefore left Denmark to be accepted with open arms by the Austrian Emperor, Rudolph. Brahe took some of his instruments and retinue with him and invited Kepler to join him in Prague. This may have been one of the most important scientific collaborations that has ever occurred.

Among the discoveries for which Kepler is famous, he is not often credited with also inventing the word *inertia* which is the cornerstone of all of physics and in fact controls all motion. In his book *The Dream*, he clearly recognized that forces of inertia, far in excess of the weight of an object, severely oppose rapid lift-off from the surface of the earth. Furthermore, he knew it was not the weight of the body, but an inertia force, which would smash the body during a hard landing on the moon. Newton later was to assign magnitudes to these inertia forces, but this detail was not yet available to the German astronomer.

Unfortunately, Kepler held on to Aristotle's belief that objects will not move unless acted upon by forces. It prevented Kepler from recognizing that bodies were able to coast along force-free forever even though a concept that represented a loose understanding of inertial motion had been discussed from as early as 1330. Many philosophers and mathematicians including Leonardo da Vinci and Kepler speculated that inertia was the very reason that Aristotle was correct in ruling that the motion of material things at all times required a force of propulsion. Unfortunately Kepler and his predecessors did not distinguish between velocity and acceleration. This subtle but absolutely crucial oversight stalled all progress in the science of inertia.

As pointed out by Max Jammer [2.7], Kepler came close to anticipating Newton's quantitative representation of the force of inertia, now known as the second law of motion. In his book, *Epitome Astronomiae Copernicanae* (1618), Kepler wrote:

> "If the matter of celestial bodies were not endowed with inertia, something similar to weight, no force would be needed for their movement from their place; the smallest motive force would suffice to impart to them an infinite velocity. Since, however, the periods of planetary revolution take up definite times, some longer and others shorter, it is clear that matter must have inertia which accounts for these differences. .... Inertia or opposition to motion is a characteristic of matter; it is stronger the greater the quantity of matter in a given volume."

In this quotation Kepler came close to defining what is meant by inertial mass and spelled out clearly that the resistance force of inertia on this mass determines the motion of a body in response to an externally applied force of a given magnitude. All that is missing in Kepler's description of inertia is a mention of the fact that the inertia force controls the acceleration of the body, not its velocity, thereby defining what is meant by acceleration and deceleration. What appeared to be a small leap, took another fifty years to be made by Isaac Newton, allowing him to formulate his laws of motion.

Chapter 2 References

[2.1]   M. Caspar, *Kepler*. London: Abelard-Schuman, 1959.
[2.2]   J. Lear, *Kepler's Dream, with full text and notes of Somnium*. Berkeley, CA: University of California Press, 1965.
[2.3]   A. Koestler, *The Sleepwalkers*. London: Penguin Books, 1964.
[2.4]   V. E. Thoren, *The Lord of Uraniborg*. Cambridge: Cambridge University Press, 1990.
[2.5]   A. P. French, *Newtonian Mechanics*, 2nd ed. New York: W.W. Norton, 1971.
[2.6]   J. A. Gade, *The Life and Times of Tycho Brahe*. Princeton, NJ: Princeton University Press, 1947.
[2.7]   M. Jammer, *The Concept of Mass*. Cambridge, MA: Harvard University Press, 1961.

Chapter 3

Free Fall

A Hardly Believable Story of Science

The free fall saga began in the fourth century B.C. with the teaching and writing of Aristotle (384-322 B.C.) in famous ancient Athens. Until that time, man's quest for knowledge was centred largely on the human body and the soul that controlled the body. Animals were subjected to the same kind of analysis. In mankind's struggle towards civilization, it was natural that its early philosophical forays concerned solely himself and his fortunes as well as misfortunes. The government of human society on earth seemed to be in the hands of the Gods, who were considered to be human, but endowed with miraculous powers. It therefore appears that for early man, everything that mattered was living and organic, the stuff of mythology and theology.

Aristotle introduced a novel idea that the world was governed by laws of nature which, in the first place, applied to lifeless inorganic matter. He called this physics from the Greek word *phusike,* meaning science of nature. Only that part which dealt with inorganic inert matter was physics. Using the word *inert* in this sense describes an objects inability to propel itself. Lifeless matter remained where it was put by external agents. However inert objects could be moving agents when they drove other inert bodies by being in contact with them. Consequently, physics became the science of matter and motion and became separated from the study of life and religion.

Aristotle was born the son of a physician in Macedonia which later became northern Greece. In the middle of his life he spent some time on the western shores of Asia Minor. This was the region where the Ionian

School of Philosophy had flourished in the previous two centuries. The founder of the Ionian School was Thales, who is now regarded as having been the first scientist of the western world. The Ionians drew attention to matter and motive forces. Aristotle took this philosophy with him to Athens where he later set up his own school to observe and investigate nature.

Figure 3.1 : Roman carving of Aristotle

As a philosopher Aristotle belonged to a privileged class which sheltered him from the turbulent sphere of Greek politics. His advantages were not based on wealth, but rather on the respect accorded to scholars. Aristotle's pride must have been wounded when, after twenty years at the Platonic Academy of Athens, and having been Plato's most prominent scholar, he was not appointed to succeed his master when he died in 347 B.C. Perhaps the disappointment inspired him to travel north, eventually finding himself at the Macedonian Court of King Philip, where he became tutor to the King's son, Alexander the Great. During this period, Aristotle learned that one of Plato's scholars had been crucified by the Persians. The martyr's last words were: "Tell my friends and companions that I have done nothing unworthy of philosophy." It is quite possible that this incident may have helped to inspire Alexander to conquer Persia not many years later. When the young Alexander came to succeed his father's throne, Aristotle returned to Athens which was still under Macedonian rule and therefore greeted his return.

Back in Athens Aristotle founded his own school located in the Lyceum, a public garden with covered walkways. *Peripatos* was the Greek word for the covered walkways in the Lyceum grounds where Aristotle taught and debated with his students, hence his teaching became known as the Peripatetic School of Philosophy.

The physics of the Peripatetics was based on four material elements: earth, water, air, and fire. The central earth was a sphere, a fact which was not fully accepted until Columbus sailed to the New World. The central and static sphere of the earth was surrounded by a layer of water and above the water Aristotle placed the spherical shell of air. All this made eminent sense. Less obvious was why Aristotle required yet another spherical shell of fire on top of the air. Apparently, this seemed natural since flames were observed to leap toward heaven. The four elements were surrounded by a number of transparent crystal spheres and each of these shells had a planet, or the sun, or the moon embedded in it. The crystal spheres were supposed to revolve around the earth to bring about the visible orbiting of our nearest celestial neighbours. The outermost shell of the world contained all the fixed stars which also revolved around the earth. This was the origin of scientific cosmology and tells us much about the intelligence of its inventor.

Aristotle handled relative motion in a unique way. He argued that one had to distinguish between natural motion and violent motion. Natural motion was the striving of matter to its natural place of rest. For instance all earth-like substances tried to go to the centre of the earth. Water fell or rose to the surface of the earth. Air drifted upward to its spherical shell and the flames of fire leapt even higher.

All other motions were violent motions. Examples were an arrow shot from a bow, galloping horses pulling a chariot or boats rowed with oars. All motions were said to be caused by motive forces. The motive force had to act as long as the motion lasted. The best understood motive force was weight. It produced the fall of a body to the ground, allowing it to repose in its natural resting place. This is how Greek philosophy arrived at the abstract concept of force as the cause of motion, very much as it is used today. It was observed that strong forces produced faster motion than weak forces. The speed of a boat could be increased by adding rowers, and a chariot could be driven faster with two horses than with one. It seemed only natural, therefore, that a heavier body, which required more muscle power to support it, should fall faster than a light body. Without experimental trials, Aristotle claimed that the rate of descent of the falling body was in proportion to its weight. It meant that if two bodies were dropped simultaneously from the same height, and one weighed twice as much as the other, the heavier one should land on earth while the lighter one was only half-way down to the ground.

Aristotle could have checked this claim by dropping two unequal pebbles from the palm of his hand. We must conclude that he never tried this, for he would have found that both landed at the same time. He must have been so firmly convinced of his theory that it did not occur to him to test it.

It was Galileo's firm adherence to experiments almost 2000 years later that eventually overthrew Aristotelian physics. However even though he did not perform experiments in the modern sense, the Greek philosopher was nevertheless an ardent observer of nature and considered it paramount for a man-made theory to agree with sensory information. He believed that observed facts "apparently" supported his philosophy and wrote [3.1]:

"I say apparently, for the actual facts are not yet sufficiently made out. Should further research ever discover them, we must yield to their guidance rather than to that of theory: for theories must be abandoned, unless their teachings tally with the indisputable results of observation."

These words might have become a canon of science, but over the millennia scientists have often maintained higher respect for their theories than for the truth embedded in empirical facts. Today the professional status and job insecurity of scientists has produced an overall unwitting reluctance for top academics to change the fundamental theories that they have taught for many years. Consequently, it takes decades, if not centuries, to adjust and replace theories that have been found to be incorrect in the laboratory.

Apart from the misconceptions of the speed of freely falling bodies, Aristotle's physics immediately encountered another problem. Why was it that a stone continued to move forward after it had left the hand of a thrower? It was felt that the moment the stone became separated from the hand it should drop to the ground, because there was no external force in contact with it to drive it on. To counter this argument, Aristotle invoked the action of a mysterious surrounding medium. It had been observed that a stone fell more slowly in water than in air. Hence it seemed that the invisible environment could influence the motion of a body. Aristotle's case was not altogether convincing. He reasoned that while the stone was still being gripped, the hand also propelled the surrounding medium. After release the medium took over the role of the contact mover and kept the stone going for a while.

This surrounding medium could not have been the air, for it was easily observed that wind did not impart a significant sideward motion to a falling stone. The inevitable conclusion was that the motive medium must be an immaterial substance of the kind much later proposed by Descartes and Maxwell, and known by the name of "ether", which had mechanical properties but was undetectable by the senses.

To delve more deeply into the science of matter and motion, we have to distinguish between *kinematics* and what constitutes *dynamics*. Consider an old-fashioned pendulum clock. Every time the pendulum is moved from one side to the other, a hand on the clock face moves one step ahead. Describing the connection between pendulum motion and pointer movement, by means of gear wheels and levers in the clock work, is kinematics. It does not involve quantitative components such as the masses of the mechanical parts, or any forces, least of all the force of inertia. Kinematics, which is the description of motion, includes the application of Aristotle's contact action physics to mechanisms, machines, and the universe. Everything that moved required something else in contact with it to move it along. There had to exist a chain of movers, like the gear wheels in the clock, one part depending on the next, and so on. In Aristotle's philosophy, looking backwards along this chain of movers, one had to assume the existence of the unmoved first mover. This mysterious entity could be the soul, or God, or some ethical principle. In fact it was Aristotle who said that love makes the world go round, implying that the unmoved first mover was supposed to be *love*.

In the clock example the unmoved first mover seems to be the earth which attracts the pendulum mass. To understand and quantify the attraction requires the modern concept of force which does not form part of kinematics. If we wish to explain how long it takes the pendulum to swing from one extreme to the other, we have to involve the mass of the pendulum and the forces of gravity and inertia which act on it. Such investigation represents the science of dynamics. Both kinematics and dynamics deal with the motion of material objects. However kinematics was a concept that was sufficient for Aristotle's physics, while quantitative dynamics was not developed until the 17th century Newtonian revolution.

At some time or another, on the covered garden walks of Aristotle's Lyceum, the philosophers must have pondered the question of what limits the speed of falling objects. If the weight force is unopposed, why do bodies not fall infinitely fast? We know it takes a stone about half a second to fall from the hand to the ground. Why does it not take one-tenth or one hundredth of a second? The answer to this question is inertia, however its cause is still being debated at the beginning of the 21st century.

Newton was clear in his own mind when he asserted that the force of inertia opposed the downward gravitational force and thereby controlled the rate of falling. Today many physicists maintain that the force of inertia is fictitious. Presumably this means that it does not exist. What then limits the speed of free fall? Pondering this problem two thousand years before Kepler mentioned the word *inertia*, Aristotle saw no alternative but to argue in favour of a medium through which bodies fell and which put up resistance to their fall. The all pervasive and undetectable ether had to step in and do the work of inertia. Ether, therefore, had to fill all space and consequently Aristotle denied the existence of a vacuum. In this way, another axiom of early Greek physics which survived until the scientific revolution in the 17th century, asserted that nature abhors a vacuum.

In the beginning of matter and motion science, Aristotle did not differentiate clearly between the velocity and acceleration of a material object. Nor was he aware that these two quantities were only measurable in a relative sense between two objects. He was apparently aware that greater force conferred more swiftness to an object. Swiftness was akin to velocity or speed, by which we mean the distance covered by, or the relative displacement of, a material entity in unit time. If a car travels at the velocity of thirty kilometres per hour (k.p.h,) over the surface of the earth, we imply that, at constant velocity, it would travel thirty kilometres in one hour. Other units of velocity are in widespread use, as for example feet per second (ft/s), miles per hour (m.p.h), and the standard scientific metric unit, meters per second (m/s). In ancient Greece distance had to be measured by builders of houses to specify the length and height of walls. Measuring rods and strings were available for this purpose. More difficult however was the determination of elapsed time, required for measuring velocity. Sun dials were known, as were water and sand clocks which were all very inaccurate by modern standards. Ordinary free fall experiments close to the surface of the earth last only for seconds or even fractions of a second. In ancient Greece, measurement of such brief intervals was practically impossible. Two thousand years later Galileo used his pulse to time the swing of a pendulum. The heart rhythm could have been employed by Aristotle, but at his time empirical science was considered less important and lagged far behind the noble pursuit of philosophy. Even today many scientists still treat accepted theory as being more valuable than a new and unexpected experimental result.

The measurement of acceleration is even more complicated than that of velocity. Acceleration is the change of velocity of an object in one unit of time. The metric unit of acceleration is meters per second, per second (m/s^2). Furthermore – and Aristotle would certainly not have known this – in inertia force calculations the acceleration of importance is that relative to the fixed stars. Aristotle launched a successful non-spiritual science of matter and motion, but it was too much for him and his generation to grasp immediately what we mean by the complex and illusive phenomenon of the inertia of matter.

What is truly incredible about Aristotle's free fall theory is that it survived two thousand years of scholastic scrutiny. At any time during this period it could have been demonstrated conclusively, with the drop of unequal pebbles or coins from the palm of the hand, that the free fall times of the bodies were not in inverse proportion to their weights, as predicted by Aristotle's theory. Scholars must have tried this simple experiment, but they simply did not wish to believe it. It was simpler to live with a lie than to upset a whole philosophy. As the ancient Greek texts were being discovered in the early European universities at the end of the dark ages, the peripatetic teaching became the scientific underpinning of Christian theology. Initially, free fall evidence was clearly not going to stand in the way of an expanding religion.

Experiment was considered to be an act of questioning God, and it became a sin to interrogate nature by physical tests. Anybody who would have tried to disprove Aristotle's free fall hypothesis in the middle ages ran the risk of being condemned by the church. The treatment of the British scholar Roger Bacon (1220-1292) is one of the best known examples of the church's persecution of an experimenter. He taught Aristotle's philosophy at the universities of Paris and Oxford and then turned his attention to experiment. He chose not to study the free fall of objects, but instead concerned himself mainly with optics and chemistry (alchemy).

In the middle of his life Roger Bacon chose to become a Franciscan friar. As such he continued his scientific investigations and wrote a scientific encyclopaedia. All of this intensified his clash with orthodox faith and Bacon gathered many opponents on the way. In a book which outlines the scholastic struggle of the church with science, White [3.2] reports that in Paris in 1278, Bacon was condemned by the General of the Franciscan order, Jerome Ascoli, who later became Pope. He was sent to prison where he remained for fourteen years. Of this experience Roger Bacon said: "Would that I had not given myself so much trouble for the love of science." Roger Bacon died in his eighties soon after his release.

It was Galileo who, at the end of the 16th century, would not allow the free fall evidence to be concealed any longer. Because of this, history would credit him with having finally overthrown Aristotle's physics. He battled the Aristotelians all his life and in the end the Christian Church found him guilty of sinful behaviour. Even though Galileo recanted his sins, his published experiments could never be hidden. In fact, they became pivotal to the scientific revolution of the 17th century. In modern times, physics is no longer constrained by the imperfections of church dogma but by the imperfections of man. Fear of damaged reputations, both of scientists and academic journals, now results in an unwillingness to constantly challenge theories, a situation that has led to a similar conservatism to that which hindered Galileo's career.

Although the free fall evidence was kept out of scholarly works for two millennia, some of the ancient Greek philosophers did criticize Aristotle's matter and motion theory. Following the early death of Alexander the Great, Aristotle left Athens, fearing persecution in a wave of anti-Macedonian retribution. His closest collaborator, Theophrastus, took over at the Lyceum in 322 B.C. and remained faithful to the teaching of his master. However, in 286 B.C. Theophrastus was succeeded by Strato who attempted to correct and extend the peripatetic physics [3.3]. He noted that falling matter was continuously accelerating and proved this to his students by observing water falls. In the upper reaches of a cascade, the water moved slowly and remained a continuous stream. Yet at the bottom of the fall the water velocity was so great that it disrupted the smooth flow. Strato also observed what we now would call the truly inertial behaviour of falling water. He pointed out that the impact with which a falling body struck the ground did not only depend on the weight of the body but on the height from which it fell, and therefore on its final velocity. Why did a ceramic cup shatter on the ground when pushed over the edge of the table and yet remain intact when dropped from a few millimetres? In only a century, Aristotle's abstract reasoning had begun to give way to experimental facts and observations of nature. Nevertheless, Strato still did not suspect that heavy and light bodies fall equally fast!

In the sixth century A.D. the subject was taken up by the Alexandrian philosopher Philoponus who became an ardent critic of Aristotle [3.3]. Philoponus directed his attention to the causes of the forces which were acting on matter. Aristotle's view had been, by and large, that the cause of the natural motion of free fall resided inside matter and could only be described as an internal force. Violent motion, such as the motion of a chariot, had an external cause (the horses) and was controlled by external forces.

Given both internal and external actions, the continued motion of a projectile still posed a problem. Was the projectile driven by an internal or an external force? It was violent motion and should therefore have been driven by an external force. Philoponus pointed out that Aristotle was wrong when he maintained that a subtile (low density) medium (ether) could both cause forces of inertial resistance as well as provide external propulsion to sustain the flight of the projectile. In this respect the Alexandrian was absolutely correct. Much later the Cartesian ether floundered for the same reason. Considerations of the flight of projectiles drove Philoponus to the conclusion that all forces were internal forces residing permanently in matter, like weight.

Here we may pause a little and examine how modern physicists view internal and external forces. Newton was quite specific that the gravitational force was a simultaneous mutual attraction of two matter particles. He could not claim that the gravitational force resided permanently in a particle of matter, as Aristotle and Philoponus had proposed because the magnitude of the attraction depended on external circumstances like the mass of the second particle and the distance between the two particles. When building force directly into matter, nature would not be able to foresee how large the force had to be. Adherents of Newton's universal gravitation have no choice but to admit that the forces of gravitation are external forces arising from the interaction of at least two particles of matter. A lone particle in the universe would display no Newtonian force of gravitation that could reside solely inside it.

Newton did not think of the force of inertia as an interaction with other particles, as Mach would later propose. For Newton, in contrast with his force of gravity, inertia was deemed an internal force which he called *vis insita*. The force of inertia was thus thought to lie dormant in a particle until it was accelerated, decelerated, or deflected from its course of motion. It was never explained how a force could permanently exist inside matter and not act all the time. In this way Newton reversed the system proposed by Aristotle. The weight force became an external force, and the inertia force, which was Aristotle's ether resistance, was made an internal force. How Newton later abandoned the *vis insita* in favour of the *vis inertia* and attributed it to an interaction with absolute space will be discussed in chapter 5.

Today we are supposed to believe in Einstein's *general relativity*. This is a field theory in which all actions are contact actions either between two particles or between a particle and a field. The gravitational attraction must then be brought about by the impact of gravitons or by the physical guidance of curved space-time. Either way gravitation remains an external force. In general relativity there exists no separate force of inertia. Einstein made the forces of inertia *equivalent* and therefore indistinguishable from gravitational forces. From this it follows that inertia forces must also be external forces.

The final chapter of this book advocates another inertia force mechanism which, like Newton's law of universal gravitation, depends on the simultaneous mutual interaction of two particles. Inertia is therefore once again an external force. Hence in the Neo-Newtonian physics, in which electric and magnetic particle interactions are also mutual and simultaneous [3.4], all internal forces have been eliminated from the macroscopic world. Nuclear forces are also said to be particle interactions mediated by packets of travelling energy. This means that they are the result of collisions and cannot be described as internal forces. In fact modern physics leaves no room for the rather medieval concept of internal forces.

It represents a total reversal from the position taken by John Philoponus. As his name suggests, he was a Christian. However, unlike his Christian successors in the middle ages, he did not accept Aristotle's matter and motion ideas. He was lucky for at his time the Roman Church had not yet developed its dogmatic scientific doctrines. Philoponus' reliance on internal forces now strikes us as alien. Yet he contributed to real scientific progress by recognizing that the ether could not simultaneously drive and retard a flying projectile. Unfortunately, this wisdom did not prevail during the re-establishment of the Aristotelian dogma in the 12th century.

Scholars of the 14th century came close to grasping the essence of inertia. Moreover, since inertia is ultimately a problem of relative motion, they also made great strides in the development of the theory of relativity. Jean Buridan (1295-1358), a French philosopher and sometimes Rector of the University of Paris, introduced a new term. This was *impetus*. If an object was given an impetus, it would continue to move as a result of the impetus. Buridan then moderated this by explaining that the otherwise continuous forward motion of a projectile was slowed down by air resistance.

Buridan's impetus has something to do with what we now call momentum (mass times velocity) of a body. In Newton's dynamics it is a mechanical impulse (force times time) which changes the momentum of the body to which the impulse is applied. Hence *impetus* and *impulse* have related meanings. They may even be identical to each other.

One of Buridan's interesting suggestions was that, at the instant of creation, God may have given each of the celestial spheres an impetus. This would have caused them to revolve around the earth forever, because far away from the earth there was no air which could retard the heavenly spheres. This sort of reasoning appeared more plausible than Aristotle's contention of the unmoved first mover, that is God, having to push the crystal spheres forever. Four hundred years later, while searching for the cause which prevented gravitational collapse of the universe, Newton considered Buridan's argument. But rather than relying on centrifugal forces to oppose the gravitational collapse, Newton thought it more likely that God lent a hand to keep the universe stable. It seems that he was forced to reach this rather unscientific conclusion mainly because he was unable to find any material objects that could be described as a cause of the centrifugal force.

Another early scholar at the University of Paris, Nicole Oresme (1320-1382), one of Buridan's students, spoke once more of the possibility, already discussed in ancient Greece, that the spherical earth was spinning about an axis. Then, without changing any of the astronomical observations, the crystal spheres could be stationary.

Oresme had a remarkable understanding of relativity, and knew already that we human beings could only observe relative motions. However Buridan had claimed that the earth was not spinning, because an arrow shot up vertically fell back vertically along the same path on which it had risen. As many other thinkers would believe in years to come, Buridan was of the opinion that a spinning earth would leave the arrow behind, so that it would not return to the point where the archer stood.

Oresme countered his professor's argument with an example of a man on a moving ship, who was unaware of the ship's forward velocity. When this man run his hand up and down the mast of the ship, he thought it was vertical motion when, in fact, the hand also moved sideways with the ship. In some way, as Oresme put it, the arrow appeared to be travelling vertically up and down, while in reality it was also moving sideways with the surface of the spinning earth. The vertical motion was an illusion, We humans could only observe the motion of one object (the arrow) with respect to another object (the earth), and we could never be aware of our motion relative to the surrounding space.

The fragments of theory developed by Strato, Philoponus, Buridan, and Oresme were finally assembled by Kepler and placed under the collective heading of *inertia*. Yet the peripatetic matter and motion philosophy was not really defeated until Galileo dispelled it with experiments which only gradually took root in the common scientific consciousness after his death. The great achievement of the Italian physicist was not that he found errors in the teaching of the medieval scholars. Others had done this before him. Galileo won the battle because he made it his business to reveal the experimental and observational truth. Then he was not satisfied with merely knowing the truth. He made it permanent by the wide publication of his results. Even with his powers of persuasion, the opposition did not relent in his life time. It was only with the next generation, that of Isaac Newton, that Galileo scored his final triumph.

The heroic Italian founder of the scientific method, Galileo Galilei (1564-1642), known by his first name, was born in colourful Tuscany, a land of olive groves, vineyards, and cypress trees between hill-top towns. He spent his first ten years in the coastal city of Pisa, where he would make scientific history only a few years later, in the gleaming white cathedral and on the separate *Campanile*, better known as the Leaning Tower of Pisa. This 55 meter tall structure, adorned with many galleries of small marble columns, was nearly one in ten out of the vertical. It was ideal for testing Aristotle's free fall thesis.

Figure 3.2 : Portrait of Galileo Galilei

Galileo's father Vincenzo married late in life and had a large family. Our hero was the oldest child. The Galileis were of aristocratic descent, but impoverished. Vincenzo was an accomplished musician, well versed in mathematics, and he wrote books on the theory of music. Unfortunately, these cultured habits only brought in a meagre financial return. He therefore looked forward to his eldest son becoming a prosperous merchant who could help to support the family. When the father discovered his son's curiosity in the laws of nature and all mechanical things, the elder Galilei had to moderate his expectations. In the end he was thankful that his son did not join a monastery while still in his teens. Instead at the age of seventeen Galileo was enrolled in the University of Pisa to study medicine and philosophy. The latter subject comprised all the science courses then available.

A passage in one of Vincenzo's books reads:

"It appears to me that they who in proof of any assertion rely simply on the weight of authority without adducing any argument in support of it, act very absurdly. I, on the contrary, wish to be allowed freely to question and freely to answer you without any sort of adulation, as well becomes those who are truly in search of the truth."

Throughout his life Galileo followed this guideline laid down by his father. The questioning he did by experiment and the answers were the experimental results. The words of Vincenzo seem to have become the basis of Galileo's scientific method. The quotation can be found in *The private life of Galileo* [3.5] which was compiled principally from his correspondence and that of his eldest daughter, Sister Maria Celeste.

Following an account by Namer [3.6], the seventeen year old Galileo rode from his home in Florence to his birthplace Pisa and there enrolled in the local university in September 1581. The freshman soon discovered the works of Archimedes, the Greek experimental genius of the century following Aristotle's. With Archimedes as his idol, the young scholar began to examine the world around him by measurement. Soon he made an unexpected discovery. At the time, no better demonstration of the effect of inertia could have been found. This was the regularity of the swing of a pendulum.

Still only eighteen years old, the observant student noticed that the period of one complete oscillation of a pendulum was constant, no matter how large or small was the amplitude of the swing. He observed this during a service in the cathedral in Pisa after the sacristan had filled an oil lamp and left it swinging on its long chain from the ceiling. Not long ago during a visit to Pisa, one of the authors, PG, could barely see the chandelier in the dimly lit cathedral. It left him wondering how anybody could have observed its motion with sufficient precision. Namer [3.6] resolved this problem when he pointed out that the chain of the lamp made a clicking noise as it moved from one extreme to the other, and it was this ticking which the young researcher heard in 1582.

By counting his pulse, Galileo convinced himself that the swing period of the lamp was constant and did not decrease as it gradually came to rest. Assuming the chain was fifty feet long, the period of one complete cycle would have been of the order of eight seconds or about eight heart beats. What was so remarkable about Galileo was that he immediately recognized the significance of his observation.

Full of excitement he returned to his god-father's house where he lodged. There he set up two lead pendulums of the same length and hung them from two different nails. His god-father, Muzio Tedaldi, helped him with the experiment. One of them drew one pendulum far aside and let it go. The other set the second pendulum in motion by drawing it only half as far aside. In spite of this difference in the width of the swing, both devices seemed to remain in perfect synchrony. The two experimenters counted one hundred swings at their respective stations and found the two pendulums were still in step with each other. Namer [3.6] describes that the young Galileo leapt with joy while the old man thought his ward was just a little crazy.

The pendulum experiment preceded Newton's *Principia* [3.7] with its theories of gravitation and inertia by ninety years. Most philosophers of this time had a fair idea of gravitation and weight forces and Kepler was just about to coin the word *inertia*. However, Galileo was the first to recognize that the pendulum bob was an object which fell repeatedly through a small distance and was therefore a controlled experiment which could be used to support or deny Aristotle's teaching of falling bodies. In his most important physics book, *Dialogues concerning two new sciences* [3.8], published first in Holland in 1618, Galileo revealed how his understanding of the pendulum dynamics grew during his life time.

He described some later experiments which again involved two pendulums of equal length, but now with unequal weights. One of them had a ball of lead and the other a ball of cork, "the former more than a hundred times heavier than the latter." Again and again he observed that when both balls, on taught strings of the same length, were released from the same height, they always arrived together at the lowest point of the pendulum swing. Because of the constraint imposed by the string, this was not a free fall experiment, but it came very close to one. It was the tension in the string which did not allow the weight to fall vertically to the ground. Nevertheless, the loss in height took the same time for the heavy and the light object. This was in conflict with the spirit of Aristotle's theory of free fall.

The pendulum is a remarkable device, which displays some of the properties of orbital motion and combines all of the forces of Newtonian mechanics; the force of gravitation, the force of inertia, and the mechanical contact force in the suspension string. Newton's theory was to help greatly in the understanding of the dynamics of the pendulum, but certain aspects relating to the force of inertia remain unclear even in modern textbooks. Galileo must have pondered that if the weight is responsible for the fall of the pendulum bob, what other force is lifting it up again during the second half of the swing? The other force had to be as strong as the weight because it returned the bob almost to the starting height. There really was no satisfactory answer to this question other than the force of inertia. The greatest difficulty for Galileo must have been to understand what determines the direction of the force of inertia. This difficulty, which even Newton did not adequately resolve, may have prevented Galileo from ever directly speaking of a force of inertia.

Figure 3.3 displays a pendulum in which the lowest point of the swing is labelled as *B* and the two highest points as *A* and *C*. Of such a diagram Galileo actually wrote [3.8]:

"This experiment leaves no room for doubts as to the truth of our supposition; for since the two arcs *AB* and *CB* are equal and similarly placed, the momentum acquired by the fall through the arc *AB* is the same as that gained by the fall through the arc *CB*; but the momentum acquired at *B*, owing to fall through *AB*, is able to lift the same body through the arc *BC*; therefore, the momentum acquired in the fall *AB* is equal to that which lifts the same body through the same arc from *B* to *C*; so, in general,

every momentum acquired by fall through an arc is equal to that
which can lift the same body through the same arc."

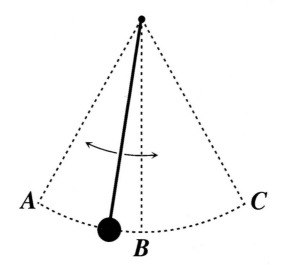

Figure 3.3 : A simple pendulum

In this quotation the English word *momentum* is the translation of
the Italian word *momento* It does not reflect the modern meaning of the
word. Today we define the momentum (mv) of a body as its mass (m)
multiplied by its velocity (v) relative to another body. The other body
has to be specified for the momentum expression to be meaningful.
Galileo was obviously not referring to the modern relativistic
momentum. His momentum was really what is now called the kinetic
energy acquired by a body due to its fall. This is given by the mass of
the body (m) multiplied by the square of the relative velocity (v^2)
divided by two, that is ($\frac{1}{2}mv^2$). Historically, the concept of energy
would not come to the forefront of physics until 200 years after
Galileo's death. Studying the pendulum from the energy perspective has
become very popular and is the only explanation found in modern
textbooks. It explains the constant periodicity without reference to
inertial forces. In a way it is an admission that scientists are still
uncomfortable with the forces of inertia.

With regard to the energy argument, all we have to remember is that a body stores gravitational energy by virtue of its height above the surface of the earth. This is called potential energy. Some of it is converted to kinetic energy during the fall of the body. This energy of motion can be used to perform useful work. For example, the grinding energy in a flour mill driven by a water wheel is derived from the kinetic energy of falling water. Whatever potential energy is not used for doing work during the fall will remain as kinetic energy. This is what happens in the pendulum as it descends from the highest point *A* to the lowest point *B*.

The cause of the acceleration of a material entity is a net force applied to it. The availability of stored energy is, by itself, not a cause of motion. For instance, unlike the concept of force, potential energy has no direction associated with it. Assume the pendulum bob sits on a table at the highest point of the swing at *C* as shown in figure 3.4. It would then possess enough gravitational energy to accelerate down to the lowest point *B*, but the force of gravity is in this case counteracted by the reaction force which the table puts up against the pendulum weight. The result is that no net external force acts on the pendulum bob which then remains sitting on the table.

Hence to fully grasp the dynamics of the pendulum we must be aware of all the forces acting on it which includes those due to inertia. An energy analysis is not sufficient to determine whether the weight in figure3.4 would move. Even though the pendulum motion looks very simple and is often called *Simple Harmonic Motion*, the force picture is quite complex. Only after Newton had discovered the magnitude and direction of the forces of inertia did the pendulum mechanism become really comprehensible. More than 300 years after the publication of Newton's dynamics, it is difficult to find the full treatment of pendulum forces, as shown in figure3.5, in textbooks of Newtonian mechanics. A detailed introduction to Newtonian dynamics appears in chapter 5. Here we will anticipate some of it to illustrate the puzzle which faced Galileo when he came ever so close to discovering the forces of inertia. To avoid the complication with air friction, it will be assumed that the pendulum swings in vacuum.

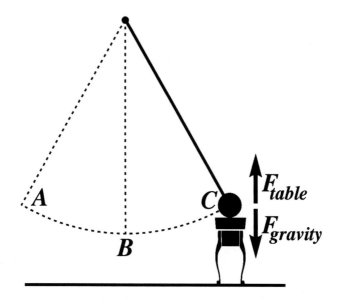

Figure 3.4 : A simple pendulum with the bob supported by a table

Readers who do not wish to follow the intricacies of the force diagram that defines the motion of a pendulum can skip the following section denoted by the dotted lines.

For those brave enough to tackle the force and acceleration vectors displayed in figure3.5, it will be seen that motion of the pendulum can be understood only if one applies the force of inertia which most importantly includes centrifugal force. As its name implies, the centrifugal force appears when an object moves in a curved trajectory. It always pushes an object radially away from the point which represents its centre of rotation.

Forces and accelerations in figure 3.5 are shown as total vectors (dotted arrows) and also shown broken up into their components in the radial (*r*) and tangential (*t*) directions (solid arrows). The force vectors have a single arrow head and the acceleration vectors display a double arrow head.

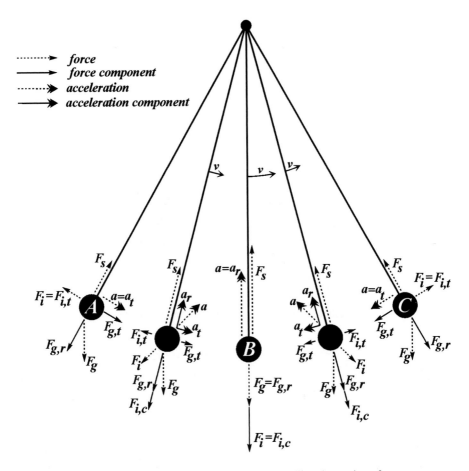

Figure 3.5 : Force and acceleration vectors controlling the motion of a
pendulum

If the pendulum is released from A, the force of gravity, F_g, pulls it
downwards toward the earth. The radial component of gravity, $F_{g,r}$, puts
tension into the string by pulling against the pivot point at the top of the
diagram. By Newton's 3rd law, the pivot point applies an equal and
opposite force on the bob, setting up a force in the string, described here
as F_s. The two applied radial forces, $F_{g,r}$ and F_s cancel each other
leading to no radial acceleration. The gravity force also has a component
in the tangential direction, $F_{g,t}$, which causes the bob to accelerate in the

direction of a_t. Since there is acceleration, a force of inertia, F_i, begins to act in the direction opposing a, which regulates the magnitude of the acceleration.

As the bob descends between A and B, it has an increasing tangential velocity and the string must be applying a force on it which is causing it to accelerate (change its direction of motion). The motion of the bob is circular and therefore the acceleration has a component in the radial direction and is shown as a_r. As in all previously discussed instances, any acceleration induces an opposing force of inertia. In examples of circular motion such as this, the radial inertial force is called the centrifugal force, $F_{i,c}$. It acts in the same direction as the radial gravity component, $F_{g,r}$, and the two of them sum together to increase the radial outward force. This increased force has to be supported by the string which causes an increase in the force applied to the bob by the string, F_s. This increase can actually stretch or even break the string and demonstrates that the centrifugal force is a very real phenomenon and is certainly not fictitious or a mathematical artefact as taught in today's textbooks. Also, as the bob descends, the tangential component of the gravity force, $F_{g,t}$, decreases, thus reducing the tangential acceleration, a_t. Consequently, the total acceleration vector, a, is constantly changing direction while the bob is in motion. The total inertial force, F_i, can therefore be seen to be a combination of its components. $F_{i,t}$ regulates the tangential acceleration and $F_{i,c}$ is the centrifugal force which contributes to the tension in the string.

When the bob is at position B, the gravity force is entirely radial and as a result, there is no tangential force or acceleration. However, the tangential velocity and therefore the radial acceleration, a_r, are at a maximum. This produces the greatest centrifugal force, $F_{i,c}$, and is the point at which the string is most likely to exceed its breaking strength. It is important to realise that the string is not just being stretched by gravity, but some other force is acting as well. This extra force is the pull of the distant universe.

As the bob continues to travel toward C, the force and acceleration vectors mirror those from the descending route. The dynamical difference is that the ever increasing tangential acceleration, a_t, is now acting to slow down the pendulum and the tangential inertial force, $F_{i,t}$ is seen to be the mechanism which in this case regulates the deceleration and is the force acting against gravity which allows the bob to gain height.

The bob eventually reaches the point, C, where it comes to rest, thus eliminating the centrifugal force and the acceleration vector becomes purely tangential. This is analogous to a ball thrown upward, slowing down to a stop at the top of its trajectory as a result of a constant downward acceleration due to gravity. At. C, the tension in the string is reduced to the radial gravity component, and the bob is in the same state as when released at A and starts to descend again.

It will be noted in figure.3.5 that at all positions along the pendulum swing, the net force on the metal sphere is zero. Opposing forces in the radial direction, that is along the string, and opposing forces in the tangential direction are in equilibrium with each other. It is as if no force at all was acting on the pendulum.

On further examination we find that the tangential force equilibrium involves a force of inertia that would not exist unless the bob was moving and specifically that it was accelerating or decelerating. We have to conclude that nature adjusts the acceleration dependent inertia force in such a way that the net force on the body is zero.

The creation of a dynamic force equilibrium of this kind, and one that arises under all circumstances, was first spelled out in 1742, fifteen years after Newton's death, by the French mathematician and philosopher Jean Le Rond d'Alembert (1717-1783). It has become known as d'Alembert's principle. Newton had already established it for the restricted case in which a gravitational force is opposed by the acceleration dependent force of inertia. However, d'Alembert's principle goes further. It asserts that when any external force acts on a body or particle, and if the material object is free to move, the resulting acceleration causes a force of inertia which will cancel the external force, resulting in a dynamic force equilibrium. Part or all of the external force is not restricted to gravity but may be of electric or magnetic origin. The qualification that the particle be *free to move* is really redundant. If the body is not free to move under the action of an external force, then that is because it is being restrained by another external force. Rather than always calling it d'Alembert's principle, in the present book we will also refer to it as *the principle of dynamic equilibrium*. This is more descriptive term for such an important physical fact which is completely ignored by many modern textbooks. Those few books which do refer to d'Alembert's principle usually imply a kind of static equilibrium, which is not the original meaning. The modern confusion stems from the fact that d'Alembert's principle clashes with Einstein's principle of equivalence, the cornerstone of his theory of General Relativity. This will be further discussed in chapter 8.

Up to this point it has been tacitly assumed that all displacements, velocities, and accelerations of the pendulum are measured relative to the surface of the earth. This is permissible because during a single swing the earth moves very little with respect to the fixed stars. To be more precise we should measure the accelerations of the pendulum bob relative to the framework of all visible stars in the Milky Way galaxy. This necessity was stipulated by Ernst Mach in the nineteenth century. The swing of a clock pendulum normally lasts a fraction of a second and at most a few seconds. During this brief period the fixed stars appear to be stationary in the sky.

The picture changes quite dramatically if we study the swinging of a long pendulum over periods of hours. The plane of the pendulum swing is then seen to rotate over the course of several hours when viewed from the surface of the earth. At the north or south pole, the plane of oscillation will rotate once every 24 hours and more slowly at lower latitudes and not at all on the equator. For a pendulum at one of the poles, it is easy to visualize that the earth revolves underneath the pendulum once per day while the plane of the pendulum swing is found to be stationary in the fixed star framework. This was first discovered in 1851 by the French physicist Leon Foucault, who also invented the gyroscope which has become the modern 3-dimensional space compass. Foucault pendulums are displayed prominently in many science museums around the world. They clearly show how the pendulum dynamics are related to the apparent motion of the fixed stars across the sky.

Pendulum experiments in particular, and free fall experiments near the surface of the earth, are almost always affected by the resistance of air. Motion through the air creates drag, which tends to reduce or limit the speed of the motion. All experimenters have been aware of this and wished they could test their pendulums in vacuum. Light bodies are more affected by air resistance than heavy objects. A leaf from an apple tree sails down to the ground far more slowly than the apple. But in vacuum both would fall together. This is demonstrated in many science museums, and to this day, 400 years after Galileo's overturning of Aristotle's physics, many people still find it disconcerting to watch a feather drop at the same speed as a stone. The effect that air has on the operation of the pendulum is that the bob, on the upswing, does not quite reach the same height as on the previous swing. So the swings become shorter and shorter until, ultimately, the pendulum comes to rest.

We could modify figure 3.5 by showing the air resistance force. In the downswing it would be in the direction of the tangential inertia force, $F_{i,t}$, and on the upswing it would oppose the tangential inertia force. Hence air resistance furnishes an external force which enters the dynamic force equilibrium. To maintain this equilibrium the inertia forces have to adjust a little. In other words, air resistance becomes another external force which causes inertia forces.

The decrease in the swing amplitude can be understood far more easily from the energy model. While the pendulum is in motion, it slightly heats up the air as it travels through it. This heat energy is lost to the pendulum system and consequently, the bob arrives at the lowest point with less kinetic energy than the potential energy that it lost. Therefore it cannot get all its gravitational energy back on the way up and stops short of the previous highest point.

Galileo [3.8] described the amplitude decrease of his lead pendulum and then added:

"Precisely the same things happen with the pendulum of cork, suspended by a string of equal length, except that a smaller number of vibrations is required to bring it to rest, since on account of its lightness it is less able to overcome the resistance of the air; nevertheless the vibrations, whether large or small, are all performed in time intervals which are not only equal among themselves, but also equal to the period of the lead pendulum."

We see that Galileo had recognized how the air resistance made the cork ball lose more of its kinetic energy on account of its lightness, or we would say today on account of its smaller inertial mass. This dependence of kinetic energy on mass is yet another manifestation of inertia. Galileo came close to discovering all aspects of inertia science, but the complete grasp of this new and difficult subject escaped him.

His greatest contribution was that he started the process of methodical experimentation. The pendulum was the first device he subjected to his empirical treatment, systematically changing the length of the string and the weight of the bob and observing the effect on the periodicity. It was the beginning of the *scientific method* which was to become enormously successful and ushered in the scientific revolution of the 17th century.

Galileo also performed a series of experiments with balls rolling down inclined planes. As the ramps became closer to vertical, he was able to develop his understanding of free fall. When the balls reached the bottom of his ramps and started to roll on a horizontal surface, he became aware that apart from the effects of friction, the balls would continue to move although not directly pushed by any external force. With these experiments, he claimed to disprove Aristotle's contention that a body can only be kept in motion as long as a force is acting on it. For this he is credited with the discovery of the phenomenon of inertia.

With regard to the actions of the forces of inertia, which are the real essence of inertia science, Galileo was less clear. He knew of course that on the upswing of the pendulum a tangential force had to exist which could lift the bob against gravity. This was similar to the qualitative understanding of inertia forces expressed by Kepler at about the same time. However it required Isaac Newton to write the magnitude and direction of inertia forces indelibly into the records of physics.

Ever since he was a student, Galileo made it his mission to disprove Aristotle's theory of matter and motion. He went about it very aggressively and was not afraid of hurting the feelings of many eminent teachers who throughout their careers had told their students that heavy bodies fall faster than light bodies in proportion to their weight. Anyone can easily test this proposition by dropping two coins of unequal weight such as a U.S. penny and a quarter. If Aristotle was correct, then if the two coins are released together at a height of 1 meter above the ground, then the penny should only be 33 cm. below the hand when the quarter strikes the ground.. Instead you will find that the two coins land together.

It is hard to believe that such a simple test was not attempted by the Italian maverick of Pisa. Yet there is no record of it. Nor can we find a record of Aristotle trying the coin drop. Is it possible that nobody carried out this test in two thousand years of scholarship and philosophy?

As it turns out, Galileo was to do something far more spectacular after he had become professor of mathematics in Pisa in 1589, at the age of twenty-five. Namer [3.6] wrote:

"Students listened openmouthed to the daring professor who gave them no text, but spoke with personal authority and begged them to turn to personal research and intelligent observation as he himself had done. ... When Galileo heard that all the other professors were expressing their doubts as to the conclusion of this insolent innovator, he took up the challenge. He solemnly invited those grave doctors and all the student body - in other words the entire university - to assist at one of his experiments. But not in the customary setting. No, that was not big enough for him. Out in the open, under the sky, in the vast Cathedral Piazza! And the academic chair clearly indicated for these experiments was the Campanile, the famous Leaning Tower.."

In the *Two New Sciences* [3.8] Galileo tells us that he made the free fall test from a great height, and found that a cannon ball of one or two hundred pounds reached the ground virtually at the same time as a musket ball weighing only half a pound. If this test referred to the Leaning Tower, then according to Aristotle the musket ball should have still been within 30 cm. of the top of the tower when the cannon ball crashed into the ground and dented the piazza.

The Aristoteleans of Pisa remained unconvinced! They continued to teach the peripatetic free fall theory. Unfortunately dogma is often held in higher regard than facts. This human failing was pointed out by Max Planck [3.9] whose experiments revealed the entirely unexpected quantum of action. He resigned himself to the inevitability that one had to wait for a new generation of scientists who could accept the unexpected findings without prejudice. For Galileo this new generation was born with Isaac Newton in the year that Galileo passed away.

A plaque on the Leaning Tower commemorates the occasion of the free fall experiment. The event was described by Vincenzo Viviani, Galileo's last student and first biographer. However in a recent biography, Ronan [3.10] claims that Viviani's story is not corroborated by other records of the time. Historians now think that the famous experiment may have been embellished by later admirers of the pioneer of the experimental method. Whatever the truth, there is no doubt that Galileo dropped weights from great heights.

As another example of how prejudice and faith overpower the discovery of contrary empirical evidence, let us look briefly at the research of Thomas Harriot (1560-1621), sometimes called the British Galileo. He was born and educated in Oxford and his primary interests were mathematics and astronomy, but later he became an avid experimenter. It was because of Harriot's application of mathematics to celestial navigation that he was chosen by Sir Walter Raleigh as his scientific companion on a journey to explore Virginia on the American continent. The two remained life long friends, and Harriot visited Sir Walter when the seafarer was incarcerated in the Tower of London for periods of time.

Very early in the 1590s, Harriot began free fall experiments with musket balls [3.11]. It must have been approximately at the same time that Galileo performed his tower experiments during his three year tenure of the mathematics chair at Pisa, between 1589 and 1592. Harriot measured the time of the fall of a bullet from 17 meters to be "more than two pulses and less than three pulses".

Later he moved his laboratory to historic Syon House on the River Thames, overlooking the Royal Botanical Garden on the opposite shore. This is the London home of the Duke of Northumberland. An ancestor of the present Duke provided Harriot with living quarters and a place to do his experiments. At Syon House, and without knowledge of Galileo's efforts, Harriot compared the fall of two different bodies. But unlike Galileo's objective of disproving Aristotle, Harriot's aim was to confirm the ancient theory. So he came to try unequal weights of lead and iron from the 13 meter high roof and recorded in his notebook "scarce sensible difference" in about twenty trials. In other words he confirmed that heavy and light bodies fall equally fast. But this was not what he wanted to find which may be the reason why he abstained from publishing the result which made Galileo famous.

The Aristotelian doctrines were defended with religious fervor throughout the Middle Ages. However in the end they were overwhelmed by contradictory empirical evidence and superseded by Newton's laws of motion and his universal theory of gravitation. However before Newton's teaching took root in the eighteenth century, inertia had become an important concept in man's understanding of motion. It played a central role in the development of the of the first post-Aristotelian cosmology, proposed by the brilliant Frenchman René Descartes. His attempt to provide a non-mathematical framework which described the universe is the subject of the next chapter.

Chapter 3 References

[3.1]   Aristotle, *On man in the universe*, L. R. Loomis (Ed.). Roslyn N.Y.: W.J. Black, 1943.

[3.2]   A. D. White, *A history of the warfare of science with theology in Christendom*. New York: George Braziller, 1955.

[3.3]   D. C. Lindberg, *The beginning of Western Science*. Chicago: University of Chicago Press, 1992.

[3.4]   P. Graneau, N. Graneau, *Newtonian electrodynamics*. Singapore: World Scientific, 1996.

[3.5]   M. Celeste, *The private life of Galileo*. London: MacMilland, 1870.

[3.6]   E. Namer, *Galileo, Searcher of the heavens*. New York: R.M. McBride, 1931.

[3.7]   I. Newton, *Principia*, F. Cajori (Ed.). Berkeley: University of California Press, 1962.

[3.8]   G. Galilei, *Dialogues concerning two new sciences*. New York: Dover, 1954.

[3.9]   M. Planck, *Scientific Autobiography and other papers*. London: Williams & Norgate, 1950.

[3.10]   C. A. Ronan, *Galileo*. New York: G.P. Putnam's, 1974.

[3.11]   J. W. Shirley, *Thomas Harriot: A biography*. Oxford: Clarendon Press, 1983.

Chapter 4

The Cartesian Interlude

A Novel Cosmology

Chronologically wedged between Kepler and Galileo on one side and Isaac Newton on the other, the illustrious French philosopher René du Perron Descartes (1596-1650) was well placed to kick the scientific revolution into high gear. He had ample mathematical ability and was adequately informed of whatever was known at his time about gravitation, inertia, and the Copernican model of the solar system. Not only did he devote himself to eliminating the myths of alchemy, astrology, and magic, he was also skeptical of some of Aristotle's ideas and criticized the blind defence of scientific dogma during the scholastic age. Blessed with the private means which enabled him to dedicate his life to study, research, and writing, mostly in tolerant Holland, he was relatively free of the obligations to church, state, and academia which have so often stifled progress in science.

His lasting legacy is the foundation of analytic geometry, which allowed equations to be more easily applied to natural events. The simplicity and utility of the Cartesian coordinate system is appreciated by every school child. With it, we can accurately define the shapes of things and also create equations to describe all motions. Every 2-dimensional or even 3-dimensional graph that we see today is comprehensible thanks to Descartes' mathematical inspiration. However, some of his other philosophical principles including his kinematical model of the cosmos and everything on earth, survived for less than a century. Nevertheless in some very influential circles, his ideas gained immediate popularity which certainly delayed the ultimate acceptance of Newton's competing philosophy.

Much of Descartes' inspiration was derived from his school days at the Jesuit Royal College de la Flèche in Anjou, founded by the progressive King Henry IV. Under the King's watchful eye and with Jesuit guidance, it offered the finest education France could provide for wealthy young boys. The competent Jesuit teachers were remarkably liberal and encouraged their pupils to examine the authoritarian views of the Roman Catholic Church. However, Descartes was disappointed that his eight years of hard study had only revealed to him his own ignorance and he felt that his pursuit of knowledge was impeded by outdated scholastic methodology.

Figure 4.1 : René Descartes

At the age of sixteen, Descartes moved to Paris determined to refresh his intellect and make his own discoveries. However after a few years, he changed course and returned to further education. Following his father's calling, he studied law at the University of Poitiers, graduating in 1616, however he never practiced the profession. Instead, he decided to retreat from the busy social and intellectual life in France and moved to Holland, where he enrolled at the military college in Breda under the command of Prince Maurice of Nassau. While pondering a mathematical challenge posted on a notice board, he met his Flemish friend Isaac Beeckman. This very well educated young doctor and exponent of the Copernican model quickly drew Descartes' attention to subjects of science such as falling bodies, liquid pressure and geometrical and mechanical problems. It seems that this friendship awakened the young lawyer's interest in physics and the philosophy of science. However before getting more deeply involved in scientific studies, he felt that he had to free his mind of conventional thinking in order to achieve something original. To this end, he enlisted in the army of the Duke of Bavaria. It gave him the opportunity to travel around Europe, and he hoped that his mathematical skills would come in useful for military engineering during the religious wars in Germany following the reformation.

In the winter from 1619 to 1620, he rented himself a room in Neuburg near Ulm, on the Danube, where his unit was stationed. He wanted to meditate beside a warm stove during the long cold months while the army was resting. He actively doubted everything he knew and tried to discover purely by thought, those concepts which could be claimed to be true with complete certainty. All of Descartes' biographers point to the significance of a dream that he had in the room at Neuburg on November 10, 1619. It was to change his scientific views for the rest of his life. During the dream he perceived all sciences to be interconnected like the links of a chain. Medicine was linked to physics, which was linked to astronomy, which was linked to mathematics, and so on. Decades later in his first major publication, *Discourse de la Méthode*, he wrote of the insight gained from his dream. From Vrooman [4.1], Descartes wrote:

"Those long chains of easy reasoning which the geometers use to arrive at their most difficult conclusions made me believe that all things which are the objects of human knowledge are singularly interdependent, and that if one will only abstain from assuming

something to be true which is not, and always follow the
necessary order in deducing one thing from another, there is
nothing so remote that one cannot reach it, nothing so hidden
that one cannot uncover it."

The chains of reasoning that underlie proofs in Euclidean geometry
can be made to hold by ensuring that every link is sound. The art of
mathematics is to check the validity of every single step in a long
argument. This is almost impossible in physics where every link would
have to be confirmed by many experiments. The most reliable theories
of physics, therefore, are those that involve the shortest chains of
reasoning. Newton's gravitation is an outstanding example of a short
chain theory and he made sure of this by applying his first rule of
reasoning [4.2]:

"We are to admit no more causes of natural things than such as
are true and sufficient to explain their appearances."

The single assumption of mutual simultaneous far-actions between
particles of matter cuts out many hypothetical steps. It eliminates the
need to define an ether or other subtle medium. It removes the present
desire to explain what flying energy (photons) is and how it knows
where to go and at what speed. Finally it does not require a further
theory regarding how this flying energy is emitted and absorbed by
matter. René Descartes employed long chains of reasoning to account
for the dynamics of bodies in motion. Some of his chains were broken
because intermediate links were found to be faulty. They were usually
hypothetical links which did not exist in the empirical Newtonian chain
of reasoning.

In 1633, after four dedicated years of writing, Descartes was ready to
declare his main ideas concerning physics and cosmology in a book
titled *Le Monde* (The World). He included justifications of the
Copernican doctrine which placed the sun in the center of the world, in
opposition to what was taught by Aristotle, the Bible, and the medieval
scholars of Christianity. Then came the news of the Pope's prosecution
of Galileo (June 23rd, 1633) for upholding the non-Ptolemaic views of

the Polish astronomer Nicolaus Copernicus. This certainly frightened Descartes and consequently *Le Monde* was delayed and only published posthumously although the principal ideas were included in Descartes' other famous publications, *Discourse de la Méthode* (1637), *Meditationes de Prima Philosophia* (1641), and *Principia Philosophiae* (1644) ([4.3] Vol.1).

Considering the lengthy Cartesian treatises on abstract philosophy, metaphysics, and ethics, it is revealed by the three essays that accompanied the *Discourse de la Méthode* that the author's expressed desire was to put knowledge to practical use. The essays were also included to show how his new method functioned when applied to the real world. The most significant of theses appendices, *La Géométrie*, was the first exposition of the principles of analytic geometry which had been maturing in his mind for 20 years and for which he will always be remembered and rightly praised. These ideas were more completely developed in his later works.

His research in the field of meteorology was revealed in *Les Meteores*, which displayed his knowledge of glaciers, winds, precipitation and provided an early optical explanation of rainbows. His most significant scientific discoveries were contained in *La Dioptrique*, in which he explained the actions of divergent and convergent lenses and was partially responsible for the light bending rule now known as Snell's law. Inspired by Galileo's well documented achievements with his telescopes, Descartes designed machines to grind non-spherical lenses and encouraged local craftsmen to build more accurate instruments. Among other goals, he wanted to investigate: "…. if there are animals on the moon?" [4.1].

At the same time Descartes vigorously criticized his Italian colleague for digressing from his line of investigation and never completely explaining his discoveries. Galileo generally did not apply the long chains of Cartesian reasoning and usually stopped when he met uncertainty. This hesitancy had also displeased Kepler who thought that his laws of planetary motion should have directed Galileo's attention to the concepts of gravitation and inertia.

It was Descartes' religious devotion which prevented him from falling out with the Church. He clearly did not harbour the rebellious ferment which Galileo had revealed, and Descartes was not sufficiently far removed from the political powers of his time who supported Aristotle. Vrooman [4.1] described an incident of which Descartes was aware

"As recently as 1624, three men had undertaken to discuss publicly, in one of the finest halls of Paris and before an audience of some thousand spectators, forty propositions that contradicted the Aristotelian school. The hall was evacuated by official order even before they would begin their speeches, their books were condemned, and the three authors were exiled. In addition it was decreed illegal to teach any doctrine that ran contrary to the "ancient authors", and the penalty for doing so was death. Such was the dominion of Aristotle, who had at his service both the Church and the state."

Threats of this kind certainly intimidated Descartes and may have subconsciously restrained his intellect from reaching its full potential. We can never know all of his private thoughts, but in his published work he was extremely careful to couch his ideas in a manner that would not contradict church dogma. Nevertheless he certainly opened up a new method for human investigation, promoting the systematic use of intelligence in order to supersede magic and superstition. His novel mathematics were a perfect example of the power of thought as a tool for discovery. However, his scientific achievements were not ground shaking principles, and his most important discoveries stemmed from his lifelong experimentation in the field of optics.

He was less careful with some of his other scientific interests. For instance he was one of the early speculators regarding the circulation of blood in the human body and the medical consequences. However even after praising William Harvey's book on circulation, he ignored the latter's experimental evidence concerning the pumping action of the heart and promoted his own theory that the heart acted solely as a heater. [4.4] His tendency to ignore pieces of experimental evidence was the reason that his dynamical and cosmological theories did not survive for very long either.

Within a generation or two, Newton's philosophy finally removed the last remnants of Aristotelian methodology from western scientific thought. Descartes and his followers held fast to some of the more questionable aspects of Greek physics. They felt that Aristotelian philosophy was sterile and not conducive to discovery, but nevertheless retained three of his four original elements of earth, air, and fire to which Descartes added a fourth, motion. The Cartesians continued to deny the existence of a vacuum and filled all space with a subtile medium which was capable of causing motion. Most emphatically, God remained the unmoved first mover.

René Descartes' approach to inertia and motion was inevitably coupled to his concept of matter. He accorded matter objective reality and recognized that the human mind can only be made aware of it by its sense organs. He argued that the senses could not deceive us because deception is repugnant to God. This however did not preclude that an observation may have several interpretations, a lesson that he learned from Galileo's discoveries. Descartes' religious belief dominated his philosophical method and he argued many times that there must be some other being, more perfect than him or his fellow men, who exists in a universe external to the human intellect.

The eminent French scientist and philosopher addressed all of the major issues of physics of his time, but he reached conclusions which differed dramatically from those of Newton, who grappled with the same problems not many years later. How could two gifted thinkers have arrived at such starkly conflicting views of the world around them?

The answer probably resides in the fact that René Descartes first formed an elaborate mental picture of the world and then tried to fit all observed facts into his grand design. Newton worked the other way around. He started with what had been observed: the fall of the apple, the rise of the tides, the orbits of the planets. Only then did he search for force laws which did no more than comply with the observations. In his speculations about the universe at large he confined himself to aspects of the cosmos with which his laws were in agreement. While Newton used particular experimental facts to guide him to the laws of nature, Descartes assumed the form of certain laws and structures of nature and then fitted experimental observations to these assumptions. Descartes is usually described as a *deductive* philosopher while Newton was an *inductive* observer of nature

Descartes' cosmology is found in his treatise, *Principia Philosophiae*. In Part II of this work, he spells out many of his assumptions regarding his view of matter and motion. Let us examine his definition of matter ([4.3] Vol.1):

"That the nature of a body consists not in weight, nor in hardness, nor colour and so on, but in extension alone."

It is hardly surprising that this definition of matter did not take root. In the mind of Descartes, body and matter were interchangeable terms, but somehow, his geometrical notion of matter and bodies completely ignored the concept of mass. He more or less held that a body was a hollow closed surface which could be said to have length, width, height, and shape, without containing any of the real stuff of the world which we now assume to be matter. Today we believe matter is a discreet amount of mass. A particle like the atom is matter. It contains a certain amount of mass. In this sense matter is the quantization of mass. Mass, and therefore all matter and bodies made of matter, are defined by their gravitational and inertial properties. It seems quite remarkable to us now that Descartes could have missed this connection.

Contrary to both Aristotelian principles as well as modern teaching, Descartes claimed that he could demonstrate the non-existence of atoms and other fundamental particles of matter which were deemed to be indivisible. He based his proof on the previously quoted assumption that matter was what we now call its volume, and any volume could always be subdivided. He went further and argued that even though one could not think of a limit to extension, the universe could not be considered infinite because only God could know such a thing. Instead he described the size of the universe as indefinite and composed of one continuous corporal substance. From this it also followed that the matter of the heavens and of the earth were one and the same, and there certainly was no room for a plurality of worlds.

The Cartesian doctrine, which held that matter was nothing more than volume, was criticized by some of his contemporaries because matter in a given volume could be rarefied. A pump could remove part of the air from a closed vessel. Some thought the amount of matter decreased inside the vessel during rarefaction. To counter these arguments, Descartes replied that during the rarefaction process pores and interstices opened up and they were immediately filled with undetectable ether. Therefore the vessel remained full at all times. So a body, which was the volume it occupied, could contain different substances, one of them being ether, but the amount of the matter enclosed by the volume was deemed to be constant. To Descartes, the weight of the body per unit volume, or its density, had nothing to do with the amount of matter the body represented.

In this way the Cartesian body became indistinguishable from the space it occupied. Yet Descartes felt that when a stone was removed from a place where it was, the extension and the shape of the stone were also removed. Hence the extent of the body, which was the essence of it, became inseparable from the body.

As cavities in a body filled up instantly with ether, it was impossible in the Cartesian body to establish a vacuum space. With nothing between them but ether, it was argued that two bodies were always touching each other. These were the assumptions on which René Descartes based his gigantic cosmic framework of ether vortices. One of his drawings of this complex system is shown in figure 4.2.

The intervention of Newtonian physics ushered in a period of 200 years in which objects, now categorized by both mass and volume, were considered to be separated by a truly empty vacuum. However, in the last 100 years, the huge empty spaces between pieces of matter in Newton's theory of universal gravitation have disappeared once again in modern physics. Instead of the Cartesian ether, we now have the subtle fluid of free energy. Gravitational energy and gravitons cannot be kept out of any vessel, regardless of how thick the walls are. Moreover, when this free energy propagates at the velocity of light, it is said to acquire mass. This energy-mass field can fill arbitrarily large spaces. Looked at in this way, modern relativistic physics is a reincarnation of the Cartesian cosmology. Following Descartes' deductive methodology, the entire edifice of modern relativistic physics is based on the Faraday / Maxwell assumption of the existence of electromagnetic fields coupled with Einstein's assumptions regarding field energy and its velocity. Based on such unobservable concepts, the deductive process has led to millions of theoretical predictions, but the foundations can never be solid. The universal cosmology taught to students today may well crumble as did Descartes' hypothetical construction.

A New System of the World, 1630-1633

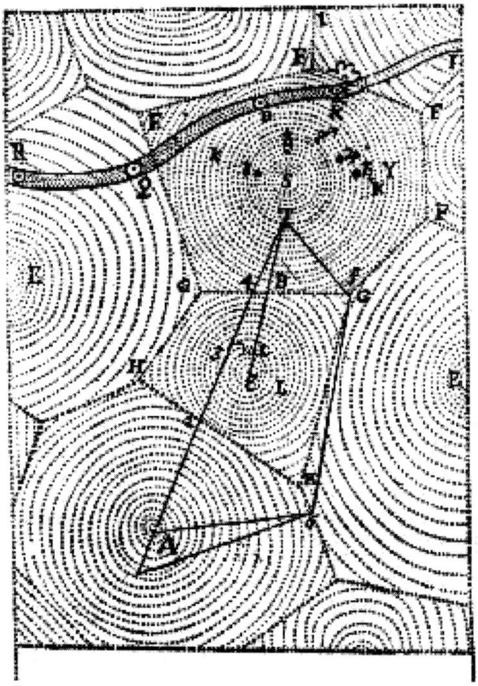

Figure 4.2 : A diagram of Cartesian ether vortices filling all space

In Descartes' philosophy, after defining matter, the next step on the way to a law of inertia was to state what was meant by motion and particularly relative motion. No distinction was yet made between velocity and acceleration. Descartes described motion as ([4.3] Vol.1):

"... the transfer of a part of matter, or of a body, from the vicinity of those that are in immediate contact with it, and which we consider at rest, to the vicinity of some others."

This immediately removed the Aristotelian distinction between natural and unnatural motions. In a further attempt to discuss relative motion, he explained that the same body could simultaneously change and not change its place. He used the example of a passenger on a ship which was traveling along a shore while standing still in his cabin, as indicated by looking at the walls, pointing out that the relative motion between his body and the seashore was not the same as the relative motion between him and the ship [4.5].

In Part II of the *Principia Philosophiae*, Descartes stated three laws of nature which were all concerned with motion and inertial effects. Haldane and Ross ([4.3] Vol.1) quoted them in English.

- "The first law of nature: that each thing as far as in it lies, continues always in the same state; and that which is once moved always continues to move."

- "The second law of nature: that all motion is of itself in a straight line; and such things which move in a circle always tend to recede from the center of the circle that they describe."

- "The third law: that a body that comes in contact with another stronger than itself, loses nothing of its movement; if it meets one less strong, it loses as much as it passes over to that body."

It is truly remarkable that early in the seventeenth century Descartes singled out inertia as playing the most prominent role in the laws of nature. While he obviously detected some aspects of the conservation of motion, he did not make it clear that inertia was related to mass. Later in the same century Newton would accord equal status to gravitation. Perhaps Newton's most profound contribution to science was to develop the concept of force which allowed him to quantify gravitation and inertia. The closest that Descartes came to discussing forces was his mention of strong and stronger bodies in the process of collisions.

In the manner of deductive science, Descartes' three laws were not bolstered with experimental facts and did not survive Newton's scrutiny. Nevertheless, they were a helpful step on the right track. The first law hints at what has become Newton's first law of motion. It tells us nothing about the meaning of *state* and does not refer to the relativity of motion. At the same time it was a major advance on Aristotle's contention that motion inevitably required the sustained action of a force.

The important aspect of Descartes' second law is the clear specification of centrifugal force. The name implies that it acts on a body moving in a circular arc and tries to make the body recede from the center of motion. The centrifugal force attempts to keep a particle moving along a straight line while it is being forced by external means to follow a curved trajectory. Consequently it is now described as one of the versions of the force of inertia.

The tendency of a body to move in a straight line is the most significant aspect of the phenomenon of inertia. Forces that try to preserve the straight line motion are forces of inertia, however no forces of inertia are acting on the body in question when it coasts along a straight line at constant velocity. It is extremely important to remember that the straight line and constant velocity have to be defined relative to other bodies. Descartes sometimes discussed motion relative to the "fixity of the stars", which he considered to be a frame that was defined by the positions of immobile matter just before God pushed the universe into motion [4.6]. This germ of an idea was a precursor to the late 18th century Machian reference frame of the fixed stars which is discussed throughout this book. Even without the relativity clause, a rigorous definition of the force of inertia as a resistance to applied forces, would have made Descartes' first law of nature unnecessary. This particular aspect of the laws of nature with respect to relative motion was obviously difficult to discern in the early years of dynamical theories for it carried over into Newton's laws of motion.

Descartes' third law begins to talk about what in modern physics is called momentum conservation. His various assertions regarding the collision of bodies do not conform with measurements. Descartes was made aware of this by a number of his contemporaries, but according to Dugas [4.5], he made excuses because bodies were not perfectly hard (of infinite rigidity). Descartes was often impatient with his critics and replied with haughty arguments. His arrogance was exemplified by the last principle that he presented in Part II of the *Principia Philosophiae* [4.7]:

"I think therefore that no principles in physics other than those which are here expounded are necessary or permissible"

There are several published volumes of objections and rebuttals ([4.3] Vol.2) which make it seem all the more remarkable that his work became so well respected. Having developed the first new cosmology since Aristotle, he looked down on his peers who tried to dismantle his splendid deductive edifice with experimental facts.

In his later years, two royal ladies were attracted to the inveterate bachelor and his glittering intellect. They both studied his philosophy and other writings. René Descartes, in gratitude for her patronage, dedicated his main work *Principia Philosophiae* to Princess Elizabeth, daughter of the King of Bohemia and Emperor of the Holy Roman Empire, at whose Court both Kepler and Brahe had served. The Bohemian Princess was in exile in Holland where Descartes lived and was eventually condemned by the magistrates for atheism.

Later the reigning Queen Christina of Sweden persuaded Descartes to come – as she put it – "to the land where bears lived among rocks and ice" [4.1]. This unorthodox queen ordered him to give her tutorials in philosophy, arranged at five o'clock in the morning. The busy courtly lifestyle and early hours apparently did not suit Descartes who even while at school gained dispensation to rise at 11 o'clock in the morning. Only a few months after his arrival in Stockholm, the Frenchman caught a severe chill and within 10 days he died on February 11, 1650, seven years after the birth of Isaac Newton in England.

What caused the enormous popularity of René Descartes during his lifetime? It was to outstrip the appeal that Newton would have on the following generation. Descartes was very clever at the art of explanation by analogy. For example, he described the motion of the planets as pieces of cork caught in a whirlpool, and light reflection was modelled as bouncing tennis balls. Moreover, Descartes was the first to describe a complete model of the universe since Aristotle, obviously an intoxicating achievement which contributed to his fame and popularity.

There is little doubt that Descartes captivated the human mind with his ether theory, the first major natural philosophy to rely on an intangible fluid to fill the spaces between matter in a plenum (completely filled) universe. This removed the need for the concept of action at a distance. Aristotle had also insisted on a plenum for the same reason, however he filled space with solid matter, his crystal spheres. Aristotle's model survived for centuries because it was difficult to disprove the existence of a transparent solid medium far away from the earth. But for the sun to push or pull the earth, the crystal substance should have been present at the surface of the earth – and it was obviously not there. The Cartesian ether seemed to overcome this problem, yet its mechanical action could not be detected in any way by the human body.

The displacement of an ether particle had to push away a neighboring particle and then another particle, and so on, until one material body could exert a force on another body. Since the world was full of ether which could not escape it, the fluid particles had to move in closed loops. This gave rise to the Cartesian ether whirlpools which were the gearwheels of the cosmic machine. Like a clockwork, everything was accomplished by contact action and kinematics. It was an ingenious invention. Descartes is purported to have exclaimed: "Give me matter and motion, and I will construct the universe". A hypothetical construction by its inventor it remained. It was a qualitative framework which did not excel in making accurate predictions which could be confirmed or denied by measurements. Galileo's experimental and mathematical science was too young and there was an insufficient amount of experimental data to seriously impede the Cartesian flight of fancy.

Sixty-three years after Descartes' death, Roger Cotes still had to forcefully battle the pro-Cartesian science establishment and defend Newton, his mentor, who had published what is now the most successful physics text of all time. Unlike those of Descartes, Newton's laws are as valid today as they were three hundred years ago. This has not stopped human minds from still dreaming about fields and flying energy and immaterial particles, filling all of space which according to Newton should be an empty vacuum. In the preface to the second edition of Newton's *Principia* [4.2], Cotes argues:

"Those who assume hypotheses as first principles of their speculations, although they afterwards proceed with the greatest accuracy from those principles, may indeed form an ingenious romance, but a romance it will still be."

He later pleads:

"Some there are who dislike the celestial physics (of Newton) because it contradicts the opinions of Descartes, and seems hardly to be reconciled with them. Let these enjoy their own opinion, but let them act fairly, and not deny the same liberty to us which they demand for themselves. Since the Newtonian philosophy appears true to us, let us have the liberty to embrace and retain it, and to follow causes proved by phenomena, rather than causes only imagined and not yet proved. The business of true philosophy is to derive the natures of things from causes truly existent, and not to inquire after those laws on which the Great Creator actually chose to found this most beautiful Frame of the World, not those by which he might have done the same, had he so pleased."

This exemplified how Cartesian physics, including his grand cosmology, became less and less relevant to the development of science. However Descartes will always demand respect for his pioneering efforts to free man from the constrictions of Aristotelian scholastic thinking, and in the process gave us one of the most powerful mathematical tools ever conceived. These stunning breakthroughs gave the human intellect the legs it required to stride into the Newtonian scientific revolution.

Chapter 4 References

[4.1]   J. R. Vrooman, *René Descartes*. New York: Putnam's, 1970.
[4.2]   I. Newton, *Principia*, F. Cajori (Ed.). Berkeley: University of California Press, 1962.
[4.3]   E. S. Haldane, G. R. T. Ross, *The Philosophical Works of Descartes*, vol. 1,2. New York: Dover, 1955.
[4.4]   J. Losee, *A Historical Introduction to the Philosophy of Science*. Oxford, UK: Oxford University Press, 1993.
[4.5]   R. Dugas, *A History of Mechanics*. New York: Dover, 1988.
[4.6]   R. Dugas, *Mechanics in the Seventeenth Century*. Neuchatel - Switzerland: Griffon, 1958.
[4.7]   J. F. Scott, *The Scientific Work of René Descartes*. London: Taylor & Francis, 1976.

Chapter 5

Newton's Force of Inertia

The Basis of Dynamics

Nearly a century after Kepler first mentioned the word *inertia*, Isaac Newton (1642-1727) gave it mathematical expression with his second law of motion. Kepler had clearly pointed out that no force would be required to move a body horizontally from one place to another if inertia did not exist. In fact this motion could then be accomplished in zero time at infinite speed. In Kepler's physics, there was no doubt that inertia had to be a force which opposed the force which caused the motion in the first place. The logic of this argument was not obvious to Newton, nor to generations of physicists that followed him. In fact, this crisis in understanding has become so extreme that many students are still taught that the force of inertia is fictitious even now at the beginning of the 21st century.

Although Newton was not really certain what caused inertia, he nevertheless correctly, but perhaps unwittingly, defined the force of inertia acting on a body, and opposing its acceleration, as being given by its mass multiplied by its acceleration. Thereby for the first time, the instantaneous acceleration of a body or a particle acquired greater importance than its velocity. The precise definition of acceleration, and particularly uniform or constant acceleration, had only recently been determined by Galileo. Armed with such a fresh and novel concept, it is uncertain when Newton first thought of the important inertia force law which became known as his second law of motion.

Figure 5.1 : Isaac Newton

Newton resolved most of the confusion in the years from 1684 to 1687 while he wrote his most important work, the *Principia* [5.1]. In order for it to be accessible to academics throughout Europe, the manuscript was written in Latin and its three books, which have become two volumes, bore the full title *Philosophiae Naturalis Principia Mathematica*. The title and the objective of the work were similar to the principal philosophical treatise of René Descartes, published earlier in the same century. Newton superseded and effectively replaced the Cartesian physics of ether contact actions with the less familiar far-actions of the *Principia*. The complete nature of interaction was however slightly confused because Newton's inertia force was some kind of contact force exerted on, or by, absolute space.

While Newton's discoveries were quickly adopted by his close colleagues, it took time for these ideas to take root in the middle of the eighteenth century after his death in 1727. In the meantime the Cartesian ether and its qualitative explanations of many observations held sway in many of the famous European universities.

Newton clearly had paid great attention to the writings of René Descartes. The first two Cartesian laws of nature, which dealt with inertia, made an impression on him but ultimately gave rise to his later confusion. Galileo had convinced everybody that, on a smooth horizontal table, balls would roll forever, were it not for the resistance of the air and friction. Moreover, when discounting resistance, the motion was along a straight line on the table. It was further observed that the ball rolled at constant velocity.

These were the experimental facts which Descartes had also expressed in his first two laws of nature. Newton later extrapolated the uniform straight-line motion to objects flying through space above the earth and between the planets. Initially Newton presumed that the behaviour of bodies was due to a continuously acting force residing inside them. He called it the *vis insita*. However as he progressed with the *Principia* he gradually came to realize that the uniform drifting of an object did not require a force at all. Newton's concepts were transformed and his enthusiasm was invigorated while working on a paper which he entitled *De motu corporum in gyrum* (On the motion of bodies in an orbit).

De motu, as it became known in abbreviated form, was composed for the British astronomer Edmund Halley (1656-1742) who is famous for having discovered and predicted the return of Halley's comet. The *Principia* might not exist but for Halley who had it published at his own expense after having persuaded Newton to write it. According to Westfall [5.2] there exist three versions of *De motu*. In the first one Newton defined his *vis insita* residing in a body as follows:

"And (I call) that by which it endeavoured to persevere in its motion in a right line the force of a body or the force inherent in a body."

It sounded rather like the Aristotelian hypothesis that a body cannot be kept in motion unless it is acted upon all the time by a motive force.

The only difference in Newton's definition was that the force acted *in* the body instead of *on* the body.

A few years later, when the *Principia* was published in 1687, it was found that Newton had dropped the idea of an inherent force required to maintain constant velocity motion. In Definition III, the definition of the force of inertia, he wrote:

> "This vis insita, or innate force of matter, is a power of resisting (not driving), by which every body, as much as in it lies, continues in its present state, whether it be of rest, or of moving uniformly forward in a right line.
>
> This force is always proportional to the body (mass) whose force it is and differs nothing from the inactivity of the mass, but in our manner of conceiving it. A body, from the inert nature of matter, is not without difficulty put out of its state of rest or motion. Upon which account this vis insita may, by a most significant name be called vis inertia or force of inactivity. But a body only exerts this force when another force, impressed upon it, endeavors to change its condition; and the exercise of this force may be considered as both resistance and impulse; it is resistance so far as the body, for maintaining its present state, opposes the force impressed; it is impulse, so far as the body, by not easily giving way to the impressed force of another, endeavors to change the state of that other. Resistance is usually ascribed to bodies at rest, and impulse to those in motion; but motion and rest, as commonly conceived, are only relatively distinguished; nor are those bodies always truly at rest, which commonly are taken to be so."

This definition hardly excels in clarity. At the same time it leaves no doubt that he understood that the force of inertia opposes the impressed motive force on the body. It is the latter which causes the change in the dynamic condition of the body. Here change means acceleration, or deceleration, or deflection from a straight line path of motion The consequence was that in the Newtonian dynamics, motion itself, at constant velocity along a right line, required no force at all. Any change in this force-free motion immediately involved two opposing forces. The impressed force brought about the change and the inertia force controlled the effect.

Newton was not always as explicit regarding the inertia force as he was in Definition III of the *Principia*. In fact when he came to discuss the second law of motion he failed to mention that it actually specified the magnitude of the force of inertia that he had already so clearly defined. More than two centuries later Einstein formulated the general theory of relativity which was meant to supersede the Newtonian dynamics. As a gross oversimplification, Einstein presumed to drop the inertia force altogether. It is now commonplace to be taught that inertia forces are fictitious. Presumably this means they do not really exist. However they are always taken into account in the design of machines and the dynamics of spacecraft. Without the centrifugal force, which is a force of inertia, the moon would fall to earth and the whole universe would collapse. The chemist's centrifuge used to separate materials of differing density simply would not work. It is shameful that modern physics textbooks deny the existence of the force of inertia simply because they do not understand its cause. The best textbooks occasionally add a note of caution. For example French [5.3] makes this point as follows.

"Once again the inertial force is "there" by every criterion we can apply (except our inability to find another physical system as its source)."

Mach was bold enough to specify that this other physical system, acting as the source of the inertial force, was the matter in the distant universe, although he did not specify a mechanism. However in Newton's mind, it seems that the inertia force acted on an isolated body and was independent of all other matter on earth and in the universe. This has proved to be unacceptable for a variety of reasons. Most fundamentally, how can the constituents of a body, its atoms, be cognizant of motion if they are in no way interacting with external matter?

Astronauts in a freely falling or freely coasting space capsule which is not firing any jets cannot tell how fast they are moving unless they interact with antennas on earth. When their radar is turned on, they detect by reflecting signals from external objects that they are in motion. When it is turned off they may easily convince themselves that they are stationary. As Newton said in Definition III : "... motion and rest are only relatively distinguished." To discover this relativity, there has to be interaction between at least two particles. On the one hand Newton was aware of the principle of relativity, which acknowledges that all we can ever observe is the velocity of one body relative to another. On the other hand he described some motion as relative to *absolute space*. He referred to such absolute motion even though he knew of no body which was at rest in this abstract entity called *absolute space*. He never succeeded in freeing himself from this residual confusion.

Newton scorned Descartes for his views on relativity which, curiously, unlike most other Cartesian arguments, have survived and are still considered to be correct. Consider once more the astronaut in his space capsule. When he turns on his rocket motor, he *feels* the inertia force, just as the driver of a motor car does when he steps on the accelerator. If the driver and the astronaut shut out the universe and just concentrate on their vehicles, there is little else to which they can attribute the felt inertia force than to acceleration relative to some fictitious absolute space. An astronaut looking out of his window and carefully observing the fixed stars – that is by letting his eyes interact with the atoms of these stars – recognizes that the force of inertia that he feels on his body always arises when his capsule accelerates, decelerates, or changes course with respect to the stars of our galaxy. This was precisely Mach's argument which will be analyzed in depth in subsequent chapters.

Should the inertia force be found to be a force of interaction with atoms in the remote universe, it could no longer be called an internal body force, as Newton had done. It would become an interaction force akin to gravity or Coulomb's law of electrostatics.

It seems there must be a close link between gravity and inertia as a consequence of the empirical fact that the gravitational mass of the body is equal to its inertial mass. The coexistence of these two mass concepts has a long philosophical history. The gravitational mass can be measured by weighing the body. The inertial mass of the same body may be determined by pushing it around on a smooth table and observing its collisions with other masses. The two masses always turn out to be equal to each other. It seems that they must be the same thing.

In that case, what is this quality of matter that we call mass? Newton directly addressed this question and said that it was the quantity of matter in a body. At other times he called it the measure of the body. To quantify anything we require units of measurement. Newton chose the unit of weight to be also the unit of mass. This was possible because the weight of the body was, by definition, proportional to the gravitational mass of the same body. Hence it was reasonable to say a body weighed one kilogram and its mass was that of one kilogram weight. The kilogram has became the widely accepted unit of mass. It could just as easily have been the pound, the ounce, or the stone. The inertial mass is of course measured in the same units. Newton defined it with his second law of motion as the ratio of an applied force to the acceleration of the body which experienced this force. Newton convinced himself by a series of experiments that the two types of mass were indistinguishable. Nevertheless, this has not precluded generations of experimenters to search for a difference. Einstein's theory of General Relativity made a subtle shift on this issue and separated the concepts of inertia and gravity, thus implying that gravitational mass and inertial mass were only miraculously equivalent but not the same thing. The most accurate experiments to date show the two quantities to be equal to at least 2.4 parts in 10^{12} [5.4].

In Newton's physics, mass is the fundamental property of matter that experiences forces. Another attribute of matter is that it occupies a certain volume of space. Matter has extent, meaning that no other matter can occupy the same space. Disregarding abstractions like *absolute space, ether, electromagnetic radiation and free energy*, it is natural that Newton's mutual forces have to act between pairs of particles of matter. There is no Newtonian mechanism available by which a particle can interact with itself, as it can in field theory. Without exception, it takes two particles to produce an interaction force.

The only particle interactions which have been observed in the last four hundred years of quantitative science are attractions and repulsions. Mutual torques, or turning moments on dipole particles, can always be resolved into attractions and repulsions between poles. Both attractions and repulsions must be thought of as single forces acting along a straight line connecting the particles. When investigating the behavior of a large collection of particles it becomes necessary to calculate the interaction force of all pairs of particles that can be formed from the collection. Attractions and repulsions experienced by an individual particle have to be added together vectorially. Each force of attraction and repulsion has a magnitude and a direction which can be added and subtracted by the special rules of vector algebra.

Newton's most important contribution to science was his clear development of the concept of force. Without it his theories of gravity and inertia could not have been formulated. Although not usually presented in this manner, his deliberations effectively culminated in three force laws. He did not spell them out in the order we will now propose here. However with the hindsight of the developments in understanding over the last 300 years, we will rank them as follows:

- The 3rd law of motion which we call the fundamental force law.
- The law of universal gravitation.
- The 2nd law of motion which defines the force of inertia.

Newton's 3rd Law – Action and Reaction

Judging by the way that Newtonian mechanics is used today by physicists and engineers, it now appears that it is Newton's 3rd law which defines the concept of the Newtonian force. In the *Principia* this law is stated thus:

"To every action there is always opposed an equal reaction: or, the mutual actions of two bodies upon each other are always equal, and directed to contrary parts."

Newton illustrated the application of this law with the example of a finger pressing a stone and the reaction force of the stone on the finger.

This refers to what are commonly called contact or local forces. When restricting the law to two particles, it implies that if an external agent pushes one of the particles against the other, and if the other is not free to move, the latter will push back on the first particle with an equal and opposite force. The two particles then, in fact, repel each other. A pair of equal and opposite contact forces are also produced when two free particles collide.

It is a little surprising that in conjunction with the enunciation of the 3rd law Newton did not immediately refer to the gravitational attraction of bodies and particles. It has to be remembered, however, that the universal law of gravitation appears much later in the *Principia*. There is nevertheless absolutely no doubt that Newton considered the 3rd law to apply to actions at a distance, and not only to contact actions. Book 3 of the *Principia* is entitled "The system of the world". It consists of both a mathematical and a non-mathematical treatment of universal gravitation. In paragraph 20 of the non-mathematical section the author says:

"For all action is mutual and (by the 3rd law of motion) makes the bodies approach one to the other, and therefore must be the same in both bodies. It is true we may consider one body as attracting, another as attracted; but this distinction is more mathematical than natural. The attraction resides really in each body towards the other, and is therefore of the same kind in both."

Later Newton explained that the same argument ought to apply to magnetic and electric interactions. In other words, the 3rd law had to be involved in the subsequently discovered laws of magnetic interaction by Michell [5.5], electric interactions by Coulomb [5.6], and current element interactions by Ampère [5.7].

Authors of many books on Newtonian mechanics have felt it necessary to improve Newton's wording of the 3rd law. An example is Barford's text [5.8]:

"The forces that two point particles exert upon each other are directed along the line joining them and are equal in magnitude but opposite in direction."

This abandons Newton's single force of attraction in favor of two separate forces. It is then no longer what he described as a single mutual interaction.

Furthermore, unlike field theories, Newton's approach did not restrict the law to point particles. Any real element of matter must have a finite non-zero volume and the distance between two finite particles is then the distance between their respective centers of gravity. Re-emphasising that the particles can only attract or repel each other makes it more obvious that the interaction is mutual and simultaneous. This feature cannot be stressed enough at a time when, in modern field theory, the mutual interaction concept has been forsaken in favour of the retarded flight of quantum particles such as photons and gravitons which transfer forces from one matter particle to another with time delays. The consequence is that field theory allows that Newton's 3rd law need not always be obeyed. However, nobody has yet observed an experimental situation in which this occurs.

The lack of any mention in the *Principia* of how the force of inertia relates to the 3rd law is a serious flaw in the internal consistency of Newton's dynamics. It was certainly not an oversight and it must have concerned Newton profoundly. In the end he claimed that the force of inertia was an interaction of one particle with absolute space. Newton did not go as far as to suggest that empty space could withstand a reaction force, however he never discussed this point for he surely knew that absolute space was unrealistic. This serious deficiency of Newtonian dynamics can be corrected with the implementation of Mach's principle to be discussed at length later in this book. It involves attractions and repulsions between an observed nearby particle and other particles in the distant universe. Only now, three centuries later, are we trying to bring Newtonian inertia forces under the umbrella of the 3rd law.

The following wording is proposed for an improved and more precise statement of Newton's 3rd law

Newton's 3rd Law Amended:

All fundamental forces of Nature are mutual attractions or repulsions of two particles of matter.

This becomes the definition of all mechanical forces in a Newtonian paradigm. An alternative name for mechanical forces is *ponderomotive* forces which can move objects through space, like pulling a cart or spinning a motor. They are measured in units, appropriately called Newtons. Another category of forces are *electromotive* forces (EMF's). They are measured in units of Volts and produce or oppose electric currents. They are non-mechanical and therefore are not covered by Newton's 3rd law. Their inappropriate name has led to a great deal of unnecessary confusion in the field of electromagnetism.

Attractions and repulsions always act along the line that represents the shortest distance between particles. This statement is so obvious that it requires no special mention in the law. Similarly, it is understood that the word *mutual* stands for *simultaneous*. In a century in which all students have been taught that forces are delayed by the flight of energy through space at the velocity of light, emphasis has to be placed on the pre-Einstein mutuality (simultaneity) of Newtonian particle pair interactions.

Furthermore, the force of attraction or repulsion is the same magnitude on both members of the particle pair. If it were not, the interaction would displace the pair of particles as a unit. This is usually described as a displacement of the combined center of mass. A similar displacement would then occur in the attraction and repulsion of particles that make up macroscopic bodies. If the forces of attraction and repulsion were not the same in both members of a particle pair, a stone lying on the ground could move itself along without being pushed or pulled by anything. Self-forces of this kind could be used to make perpetual motion machines and are often called bootstrap forces for if they were real, one would be able to lift oneself out of quicksand by pulling on your own bootstraps. Only the remarkable Baron Munchausen of fairy tale fame has managed this impossible feat. This is very strong experimental confirmation of Newton's 3rd law.

The magnitude of the mutual attraction or repulsion depends on the nature of both particles. In gravitation, for example, it is proportional to the two masses. This implies that the particles must be aware of each others' mass. The exchange of information between the particles, limited to travelling at the velocity of light, could not achieve this simultaneous awareness between all particles in the possibly infinite universe. The strength of the interaction force is also a function of the distance of separation of the two particles. This again requires a simultaneous awareness of each other. The human mind apparently finds this mutual awareness between pieces of dead matter difficult to accept. Einstein referred to it as "spooky action at a distance". But is it really more ghostly than the invisible, intangible, and altogether undetectable flying forces of non-material energy particles like the photon? Physics and physicists nevertheless have to choose one undetectable mechanism over the other and really should know better than to use 'spookiness' as a criterion for deciding the validity of physical theory. Many experiments in electrodynamics and quantum mechanics favour explanations employing far-actions over delayed flying forces [5.9]. Inertia, particularly, would forever remain a phenomena without a physical cause if physics could not be based on the principle of far-actions.

It is significant that the 3rd law contains no reference to space, be it empty or filled with ether or free energy. Nor does it concern the flow of time. Only elements of matter are involved and their relative positions. The 3rd law ensures a natural and believable relativity, known as Galilean relativity. It is independent of the whereabouts and motions of observers and observing instruments. This is the hallmark of Galilean relativity as opposed to Einstein's observer based relativity in which the magnitude of forces depends on relative velocities with respect to the speed of light. With Galilean relativity the laws of nature appear to be the same everywhere, no matter who makes the measurements to determine them as long as they take into account their own acceleration which they can feel by the inertial forces acting on them.

René Descartes referred to this natural relativity as the only logical system of philosophy. Bizarrely, and probably out of fierce rivalry with his French counterpart, Newton disagreed. He did not recognize that his own all important definition of mutual force by the 3rd law, but not the absolute space of his inertia model, led to the natural Cartesian relativity. This confusion remained in Newton's mind up to the end of his life. It is however easily understandable how someone who turned the tables on two millennia of Aristotelian physics, and then demolished the popular ether universe of Descartes, might have missed a few cobwebs of his prior beliefs. Those remaining relics of absolute space and time, and their connection to Newton's force of inertia, were nevertheless bound to come under attack sooner or later.

Newton's Universal Law of Gravitation

Newton's universal law of gravitation became the leading practical application of his 3rd law. This marvellous mathematical statement that united earthly events with the motions of the heavens brought Newton more fame than his laws of dynamics. In the years after Newton's death his 3rd law spawned three more fundamental universal force laws. All of these were inverse square laws. However, unlike gravitation, they also involved repulsions. The first of them was Michell's law of the interaction of magnetic poles published in 1750 [5.5]. He had discovered that like poles repel and unlike poles attract and the strength of the mutual force was related by the inverse square of the distance between them. Twenty-five years later this was followed by the analogous and much more famous Coulomb's law of electrostatics [5.6], describing the attraction and repulsion of electrically charged bodies and particles. In 1822 Ampère [5.7] discovered the Newtonian inverse square interaction force between metallic current elements in electrical circuits. All four of the simultaneous mutual interaction laws were of empirical origin. They were derived from simple experimental observations without recourse to any theory and for this reason they will be valid forever unless Nature changes its laws from time to time.

The reason that we have ranked the law of gravitation ahead of the force of inertia is simply that for gravitation, Newton for the first time made the bold assumption that an interaction force could span any distance. All the astronomical evidence at our disposal still suggests that this is true. To proclaim that gravitational attractions were truly universal and reached the furthest corners of the cosmos was a most courageous act coming, as it did, after a two thousand year reign of Aristotle's contact action dogma. Universality almost certainly also applies to the force of inertia. This is difficult to grasp, and would be rejected by all, were it not for the prior demonstration of universal gravitation.

It is uncertain when Newton convinced himself of the validity of the law of gravitation and the related formula. Late in life, as an octogenarian, he told his much younger friend William Stukeley [5.10] that the idea came to him when he observed the fall of an apple at home in the Lincolnshire countryside. Even today there stands an apple tree in front of Woolsthorpe Manor, the farm house in which Newton was born. The small cottage is now a museum that contains many artefacts of Newton's life including some of his scientific sketches on the walls. Stukeley's story traces the discovery of the force of gravitation back to a two-year period from 1664 to 1666 which Westfall [5.2] has described as Newton's *anni mirabiles*. For much of this time Newton was staying with his mother at Colsterworth, Lincolnshire, in the Midlands of England. He was then in his early twenties and the black plague had closed the doors of Cambridge University where he had studied since 1661. This return to his childhood home freed him of the burdens of university life and undoubtedly proved a source of inspiration. Escaping from city life to the green landscape probably brought back the dreams of his boyhood. There he was to ponder the great unsolved questions of science and mathematics.

The name, *Woolsthorpe Manor*, suggests a more substantial establishment than history relates. Nevertheless, with more than 200 sheep to his name, it was several times as large as neighboring farms. The Lord of the Manor had few privileges in the village of Colsterworth. Isaac's childhood was loveless and full of anger. His father had passed away a few months before his birth on Christmas Day 1642, the same year in which Galileo died. As Westfall [5.2] put it: "Prior to Isaac, the Newton family was wholly without distinction and wholly without learning."

When the boy was three years old, Isaac's mother married again, moved to another village in Lincolnshire, and left her child in Woolsthorpe Manor in the care of his grandparents. Not much love was lost between the three who stayed behind on the farm. Perhaps Isaac was simply not a loveable little boy for he made few friends and seems to have been quarrelsome. His early years coincided with the closing stages of the English Civil War during which the social and political structure of the country was in turmoil with open divisions in every town and village. Quite probably, Newton's escape into abstract thought was precisely the expected reaction from a highly intelligent youth, unable to reconcile the mad world in which he lived.

Figure 5.2 : Woolsthorpe Manor and apple tree

When Newton was ten years old his mother returned, following the death of her second husband. With her came a half-brother and two half-sisters. No close relationships developed, yet his mother cared enough to ensure that Isaac was educated. He first went to village schools and from the age of twelve he moved to the Grammar School of the nearby market town of Grantham. Isaac was an introvert and disliked by his peers for his mental agility. His performance in class lapsed at times because he entertained himself with the building of wooden models, sundials, and dripping water clocks. However on several occasions after being reprimanded by his master, Isaac would effortlessly return back to the head of the class.

During school terms he boarded in the Grantham apothecary. Possibly the only romantic involvement he ever formed was with the pharmacist's daughter. He is said to have made dolls houses for her.

At the Grammar School, Newton learnt to read and write. Most important were his lessons in Latin which stood him in good stead in his professional life. He also studied the Bible and may have read other books. However mathematics was not on the Grantham curriculum. When he reached his seventeenth birthday his mother thought he had received enough education and should prepare himself to become Lord of the Manor.

Luckily for the history of science, his agricultural apprenticeship was not a success. He set up waterwheels in the ditches of the farm while he was supposed to be minding sheep and pigs. With his attention elsewhere, the animals strayed in the neighbor's cornfield. For this he was fined and his mother had to pay damages.

Fortune saved the genius, as it would several more times in his life. The headmaster in Grantham and a caring uncle convinced Isaac's mother that her son should go to university. He was forthwith returned to Grantham Grammar School to be prepared for entry into Trinity College. In June 1961 he went to Cambridge to study mathematics. He might just as easily have been asked to do medicine, or law, or prepare himself for a priesthood. At this important crossroads, his natural curiosity of the physical world steered him toward science where he would excel as nobody had before or since.

Less than four years later Isaac Newton would return to Colsterworth for his *anni mirabiles* to lay the foundation for his differential and integral calculus and much more. He began to sense that he was destined to unravel more secrets of nature. He must have felt this when, one day, he sat in a contemplative mood under the apple tree and suddenly understood that the falling apple was attracted to the center of the earth.

His uncle William Ayscough, a Cambridge educated minister, who had encouraged him to go to university, died in 1669, three years after

Isaac had returned to Cambridge after the plague had subsided. Uncle Ayscough was one of the few persons who cherished the bright nephew and in his will he left a sum of money to his goddaughter Frances Newton. There is no hint in the church records of Colsterworth and neighboring villages who the little girl Frances Newton was. This fact was unearthed by a close personal friend of ours, Kenneth Baird.

Being a Cambridge historian and a resident of Colsterworth, Ken became involved with Woolsthorpe Manor and researched the family history of the Newtons and Ayscoughs. Not averse to mixing humanity with science, he pointed out that, on the basis of dates and relationships, it was not out of the question that Frances was Isaac's illegitimate child. In a note to the Royal Society of London, Baird [5.11] wrote:

> "It is thin evidence that Isaac Newton was not the virgin he claimed to be when he was an old man, and thin evidence against his apparent antipathy for romantic attachments to women. Nevertheless it is not altogether negligible evidence, and it deserves to be put on record."

Nobody knows what happened to the girl from the apothecary!

Newton's personal life was certainly not his strong suit, however the unravelling of the astute but muddled efforts of his scientific predecessors was how we should remember him. Gravitational attraction was by no means a new idea. Kepler had maintained that the planets were held in their orbits by attraction to the sun. Others associated the tides with lunar attraction. What may have occurred to Newton under the apple tree was that the three phenomena of free fall, tidal flows, and planetary orbiting were the result of a single physical force which could be described by a mathematical law, and that attractive forces were *universal*. The idea of action at a distance may at that time have also crossed his mind, but it was many years before it took root.

To have a flash of insight into the workings of nature must have been an exhilarating experience, difficult to forget in a long life. As it turned out, the ultimate emergence of the attraction formula was a twenty year long grind. It was not settled until the publication of the *Principia* in 1687. The complications stemmed from the fact that the force of gravity could not be derived in isolation of the rest of physics. Most of all, it was intertwined with the science of dynamics started by Galileo, analyzed by Descartes, and left incomplete for the lack of a force of inertia. Newton could see that gravity and inertia forces were intimately coupled to each other in making planets orbit around the sun. Both were the result of the very essence of matter – its mass. It was, therefore, not surprising that the formulae of the two forces emerged simultaneously. As soon as they had revealed themselves, Newton was ready to publish.

Between 1666 and 1684, Newton had made little headway with his "mechanical philosophy", as he called it. Most of this time was devoted to alchemy and theology, and his duties as Lucasian Professor of Mathematics. His biographer Westfall [5.2] made this point.

"As it appears to me, Newton's philosophy of nature underwent a profound conversion in 1679-80 under the combined influence of alchemy and the cosmic problem of orbital mechanics, two unlikely partners which made common cause on the issue of action at a distance. ... Henceforth, the ultimate agent of nature would be for him a force acting between particles rather than a moving particle itself ..."

It was the all pervasive nature of the 3rd law which revolutionized Newton's thinking. Simultaneous attraction became to him ever less objectionable and less occult. He believed also that solid bodies were held together by the attraction between their particles. Such an important and tangible force simply had to be real. Newton was lucky that in England he enjoyed relative academic freedom and was able to air his views without the fear of persecution that had restrained many of his continental predecessors.

Two comets appeared in the early 1680s. The second one was a visit by the familiar object, now known as Halley's comet. An argument developed between the Astronomer Royal, John Flamsteed, and Newton. Interestingly, both men believed that, since comets came from the outside of the solar system, their paths were not prescribed by the same laws of physics that governed planetary orbits around the sun. It seems Newton had still not yet fully recognized the universal nature of gravity.

In 1684 fortune would shine once more on Isaac Newton and it led to his greatest success. The astronomer Edmund Halley visited him in Cambridge and asked what he thought the shape of the orbit would be of a planet encircling the sun if the planet was attracted by a force proportional to the inverse square of the distance? An ellipse, answered Newton without hesitation. Halley had come to the same conclusion. What is more, it agreed with Kepler's observation based laws. This conversation with Halley, and the latter's support and determination, directly led to three years of phrenetic intellectual activity which culminated in the publication of the *Principia*.

Within months of his visit, Halley received the paper, *De motu*, which contained Newton's mathematical demonstration of the elliptic orbit of a planet with the sun as one focus of the ellipse. In this paper Newton used the new term *centripetal force*. By this he meant a force seeking the center of the orbit. Referred to planetary motion, this was the gravitational attraction. The Dutch physicist Christian Huygens (1629-1695) had coined another new term, *centrifugal force*, which stood for fleeing from the center. The centripetal and the centrifugal force on the planet opposed each other with equal strength. The centrifugal force is a force of inertia because it opposes the deflection, or inward acceleration, of any orbiting body. Gravity tries to accelerate a planet toward the sun and the inertia force opposes this radial inward acceleration. Newton recognized that the stability of planetary motion required the two forces to be exactly equal to each other. In modern inertia theory this will be called the condition of dynamic equilibrium. It is a constant equilibrium condition and it defines the magnitude of the inertia force.

Apparently it was Huygens who first proposed the formula for the centrifugal force. With it Newton, and also the British physicist Robert Hook, arrived at the inverse square law of gravitational attraction via Kepler's formulas of orbital kinematics. Hook claimed priority. It is not uncommon that independent investigators reason along parallel lines and arrive at similar conclusions, often at the same time. History has awarded Newton the credit for the universal law of gravitation which was fair because Newton justified the law in much greater detail. He compared his law with many astronomical observations, the behavior of the tides, and laboratory experiments. Beyond this, Newton embedded the law in a new mechanical philosophy that revolutionized physics.

There is no concise statement of the law of gravitation in the *Principia*. The modern formulation of the law condensed from Newton's works can be written as;

Newton's Universal Law of Gravitation

Every particle of matter attracts every other particle of matter with a force varying directly as the product of their masses and inversely as the square of the distance between them.

It must be remembered that attraction always stands for a simultaneous interaction between a pair of particles. It is defined by Newton's 3rd law. The philosophical arguments and the experimental facts which support a physics paradigm based on simultaneous far-actions at the beginning of the 21st century have been more fully discussed in an earlier book of ours [5.9]. Every particle attracting every other particle simultaneously leads to an interconnected universe. This differs from the disjointed world of conventional field physics where the field energy required to cause forces is assumed to travel with the speed of light from one particle to another. The principle of instantaneous interconnectedness is essential in the Machian derivation of the force of inertia discussed throughout this book. Although the individual far-actions between particles separated by cosmic distances will be extremely weak, every speck of matter of the Newtonian universe is aware of all other matter and instantly senses and reacts to any changes in the distribution of that matter.

Newton's 2nd Law – The Force of Inertia

The derivation of Newton's second law of motion, which is the inertia force law, proved intellectually more challenging than that of the other two force laws. Newton's world view did not conform with the 3rd law for it discussed the action of an inertia force on a single particle without mention of its cause or reaction. Newton must have been aware of this inconsistency but he expressed no concern. As late as 1684 he still had not purged his mind of the *vis insita* concept. However, this must have changed in the following years for when the *Principia* was published in 1687 he stated his 1st law of motion without any reference to an inside force.

"Every body continues in its state of rest, or of uniform motion in a right line, unless it is compelled to change that state by forces impressed upon it."

No longer was there any need for the *vis insita* to maintain uniform motion. In fact, there was really no need for the first law itself. It involved no forces other than the reference to the external force required to change the motion of the body. If the inertial behavior of matter, that is its resistance to acceleration, required a force of resisting, then this would have to be determined by another law which became known as the 2nd law.

Unable to free his mind of this Cartesian hangover, Newton was wary of such a drastic change in his outlook. This ambiguity was reflected in his formulation of the 2nd law of motion.

"The change of motion is proportional to the motive force impressed, and is made in the direction of the right line in which that force is impressed."

The impressed force is, of course, not the force of inertia, but rather is the action of an identifiable source; the push of a finger or the gravitational pull of a star. As spelled out in Definition III in the *Principia*, the force of inertia opposes the impressed force. However, this is not clear from Newton's wording of the 2nd law. This confusion has continued to exist throughout history up to the present time. The common view now is that the second law defines the concept of force, and the force of inertia is *fictitious*. In other words the force of inertia can be made to disappear by choosing an appropriate frame of reference. Some clarification emerged in the nineteenth century with the writings of Mach. He pointed out that we have only one true inertial frame of reference and that is the frame of the fixed stars which is the only one in which momentum is always conserved. In this frame of reference the force of inertia is very real indeed.

Once the 3rd law has been made precise as we have done, there is no need for another definition of force. The 2nd law then can assume the role for which we believe that it was designed. With this in mind, Newton's 2nd law may be amended to specifically define the force of inertia and would read:

Newton's 2nd Law Amended

The magnitude of the force of inertia, F_i, acting on a particle is equal to the product of the mass of the particle and its acceleration relative to the fixed stars ($F_i = -ma$). It is equal and opposite to the externally impressed force, F_e, ensuring that ($F_e + F_i = 0$).

The principle of dynamic equilibrium follows from this form of the law. It asserts that a particle will accelerate just enough so that the opposing force of inertia is precisely equal and opposite to the force of propulsion. Therefore acceleration is controlled by the force of inertia. Without it a body would respond to an impressed force with infinite acceleration and a stable universe would not be possible.

In the case of free fall, the force of gravity and the force of inertia oppose and cancel each other. It suggests that the force pair can be dropped from the theory of gravitation. This is precisely what Einstein did in his general theory of relativity which is a theory without forces in which gravitational motion is caused by curved space instead. However, the general theory of relativity has the problem that the external force may not be due to gravity. For instance, it could be a magnetic force rather than a force of interacting masses. In that case, the force of inertia still needs to exist in order to control the relevant particle accelerations. However, according to general relativity the force of inertia is not real. This is one of the reasons why the theory of general relativity has not been extended to embrace electromagnetism. The important conclusion is that in Newtonian mechanics we cannot ignore the impressed force just because it is equal and opposite to the force of inertia, for the acceleration of the body would simply not occur if the external force was ignored. The force of inertia is never a cause of motion, but is always a passive consequence of acceleration.

In our amended version of Newton's 2nd law above, the clause *acceleration relative to the fixed stars* has been inserted. Newton however believed that it was *acceleration relative to absolute space*. He never stated that this abstract absolute space was really the same space in which the firmament of stars appeared to be stationary. If at times he did recognize this possibility, he probably saw no way in which the fixed stars could produce a causal interaction with the earth which gave rise to the force of inertia. The body of 17th century astronomical knowledge demonstrated an uneven distribution of stellar bodies which probably convinced him that his proposed force of gravitational attraction could not be related to inertia. It would have predicted that the magnitude of the force of inertia was dependent on the direction of an external force. However, all experiments have indicated that the inertia force is isotropic, meaning it is the same in all directions.

When in his version of the 2nd law Newton said: "The change of motion (of a body) is proportional to the motive force impressed"; he makes no mention of what other body the former is moving with respect to. Nor does he explicitly state that he meant motion relative to absolute space. Nevertheless, from the various examples that he cited, it is clear that he referred to change of motion relative to absolute space. It is not inconsistent with the space in which the fixed stars were at rest. However from later comments we gather that he thought that absolute space might not be detectable with our senses. In Newton's mind the framework of the fixed stars was not absolute space because man could see the fixed stars and probably for theological reasons, absolute space had to remain undetectable.

The fact that Newton did not couple his 2nd law to acceleration relative to the earth demonstrates his profound understanding. A terrestrial frame of reference would have worked for all falling objects observed near the surface of the earth. According to the 2nd law, it would also have worked for a cart or sled travelling on a horizontal road or up and down hills or a cannon ball in flight. The earth is simply so much more massive than any of the objects in our every day existence that we simply do not notice that its motion is ever so slightly affected by every acceleration that it causes. So why did Newton feel he had to call upon absolute space?

The reason he gave was the centrifugal forces on bodies revolving around each other in deep space far removed from the earth. Remembering that by true motions Newton meant absolute motion. Relative motions between two bodies he called apparent motion. He described his view in the famous Scholium on absolute space and time in the *Principia* as follows.

> "It is indeed a matter of great difficulty to discover, and effectually to distinguish, the true motions of particular bodies from the apparent; because the parts of that immovable space, in which those motions are performed, do by no means come under the observation of our senses. Yet the thing is not altogether desperate: for we have some arguments to guide us, partly from the apparent motions, which are the differences of the true motions, partly from the forces, which are the causes and effects of the true motions. For instance, if two globes, kept at one from the other by means of a cord that connects them, were revolved about their common center of gravity, we might, from the tension of the cord, discover the endeavor of the globes to recede from the axis of their motion, and from thence we might compute the quantity of their circular motions."

In other words, the tension in the cord convinced Newton that absolute motion is taking place and this occurs independently of any motion relative to the earth. Near the end of the Scholium Newton almost admits that this absolute motion could be relative motion with respect to the fixed stars. His actual words were:

> "But now, if in that (absolute) space some remote bodies were placed that kept always a given position one to another, as the fixed stars do in our regions, we could not indeed determine

from the relative translations of the globes among those bodies, whether the motions did belong to the globes or to the bodies."

It is a curious fact that the Newtonian dynamics is perfectly compatible with the principle of relativity, in spite of Newton's insistence of absolute space being the cause of the force of inertia. The ordinary principle of relativity asserts that the motion of a body cannot be measured other than by comparing its positions with those of a second body. Space as an entity dissociated from matter plays no part in relative motion. This again describes Galilean relativity for it was Galileo who first discussed that a sailor travelling at constant velocity on smooth sea inside his boat, cannot determine whether he is moving or stationary without looking out of the window toward the shore. In other words, no motion is absolute. Our definition of force as pronounced in the amended Newtonian 3rd law relates all forces, including the force of inertia, to pairs of particles and their distance apart, that is to their relative positions. Therefore there can be no conflict between our revised Newtonian dynamics and Galilean relativity.

The interaction of mass with absolute space was a blemish of Newton's theory of dynamics. Mach started the removal of this flaw in the nineteenth century and as will be demonstrated in chapter 12, it now appears very likely that the forces of inertia are caused by the distant matter in the universe and comply with Newton's 3rd law. This seemingly small theoretical discovery has huge philosophical ramifications which may force a return to a physics paradigm regulated by instantaneous Newtonian forces between pairs of particles at any distance of separation. We could then free ourselves of the unverifiable notions of field energy and spacetime. This bold new conception will be developed further at the end of the book.

However returning to the historical development of the subject, in the 1680s, while Isaac Newton wrote his *Principia* in England, his chief rival on the continent was the Dutchman Christian Huygens. He anticipated Newton with the notion of *acceleration due to gravity* and with the formula for the centrifugal force. As Barbour [5.12] has pointed out, Huygens could easily have won the greatest prize, the fame for creating the science of dynamics. Unfortunately history rarely relates the political side of science by which recognition is rightly or wrongly awarded. However, the man who is clearly deemed to have compiled and created the most complete and successful dynamical model was Newton. Réne Descartes died in 1650 and Huygens was his very able successor. In the end it was Huygens' adherence to Cartesian principles which robbed the Dutchman of his laurels. He apparently could not reconcile himself to the concept of forces between separated objects, action at a distance.

Manifestations of the Force of Inertia

As early as 1659, before Newton's *anni mirabilis*, Huygens proposed the formula for the centrifugal force. This is the force which prevents the orbiting planets from falling into the sun. It counteracts gravitational attraction and is an example of the force of inertia. Without centrifugal forces the universe would collapse, a potential outcome that has been a major consideration of all cosmologies since that time. In modern science it is argued that gravity and inertia are opposing but nevertheless unreal properties of matter. This is quite opposed to the view held by the philosophers of the 17th century scientific revolution. They believed that the forces of gravity and inertia were real, of differing origin and counteracted each other, thereby ensuring the relative stability of the universe.

The effect of inertia is usually described in three guises; resistance to linear acceleration, the centrifugal force and the Coriolis force. Then as an afterthought it is usually added that the centrifugal and Coriolis forces are illusory and only occur as a result of observing events from a particular rotating vantage point. While all three of these effects are ascribed to inertia, it is a peculiarity of modern physics that one inertial effect should be considered real and the other two fiction.

It is unanimously considered to be true that if the same force is applied by a common source to two objects, the one with the greater mass will accelerate more slowly than the one with the lighter mass. This is of course Newton's 2nd law discussed earlier in this chapter. In this respect we often speak of the greater inertia of a more massive body. It is the reason that a large vehicle requires a large and powerful engine to achieve the same acceleration as a toy car with a tiny motor. This effect can be viewed as occurring either as the result of a force of inertia which opposes acceleration against the backdrop of the fixed stars or more conventionally simply as compliance with the traditional treatment of Newton's 2nd law.

This issue was muddied by Einstein's theory of General Relativity. Using the fact that objects under the effect of gravity fall with an acceleration independent of their mass, he assumed that there was no force of gravity and thus there could also no longer be an inertial force resisting the acceleration for it would then be the only force acting on the object. By this pure assumption, the force of inertia was eliminated and inertia simply became the unexplained consequence of Newton's 2nd law. Einstein decided to ignore the fact that in the case of gravity and inertia, both of these forces depend on mass and only in this situation are the accelerations independent of mass. In the case of a magnetic force or even a direct mechanical push combined with the consequent inertial force on a body, this cancellation does not occur and its acceleration definitely does depend on its mass. Einstein's presumptuous removal of both the force of gravity and the force of inertia was seen at the time as a liberating revolution in which the local structure of spacetime around us was considered a more realistic controller of dynamical effects than the distant far-actions of Newtonian physics. The vehemence of the rejection of the existence of inertial forces is quite remarkable and can still be detected in the adamant tone of mechanics textbooks written today when addressing these concepts.

However hard it may be to determine that inertia is a true force when it acts directly in opposition to an object's motion, there is much more definite proof of its existence when it acts in a direction not in the same line as the instantaneous velocity. The most striking example occurs when an object is subjected to a force that is perpendicular to its velocity of motion. This situation occurs when an object is moving in an arc of a circle or ellipse and the consequent inertial force is called the centrifugal force. It was named and first discussed by Huygens who arrived at his quantitative specification of the force from a study of Galileo's experiments with falling, sliding, and rolling bodies. It was a triumph which revealed the analytical power of Huygens' intellect. What Newton realised, and Huygens missed, was that the centrifugal force always acted in opposition to the inverse square gravitational attraction between orbiting objects. The two forces together led to an explanation of Kepler's laws and the known orbits of the planets. This final step was possible as a result of Newton's adherence to his 3rd law embodied in gravitational attraction. Huygens adamantly refused to believe in this simultaneous attraction and in his mind retained the ether contact actions of Aristotle and Descartes.

Figure 5.3: Christian Huygens

It is important to understand what determines the size of the centrifugal force. Since it is based on the amount of acceleration required to make a body move in a curved path, it can be deduced that the magnitude of the centripetal and centrifugal force on an orbiting object is proportional to the square of its velocity divided by the radius of the curvature of its path at any time. Its velocity will depend on its previous history, but if the consequent radius of curvature of the path happens to correspond to its distance from the body providing the centripetal force, then the particle is in a stable orbit. This is true for objects acting under the effect of gravity, but it equally applies to the action of spinning a stone around one's head on a piece of string. In the case of gravity, the balance is quite delicate because the force of attraction is distance dependent, meaning that for a particular velocity, there is only one possible distance of separation that defines a stable orbit. If we are simply swinging an object around our head on a tether, we know from our childhood experiences that the system is not stable until we have given it some rotational velocity so that there is some centrifugal force. We then provide an equal amount of inward centripetal force with our arm muscles to keep the rotation steady. If we swing the stone faster, then due to the fixed length of the string, the centrifugal force increases, thus requiring us to use more strength in our arm muscles to supply the required centripetal force to maintain the stable orbit.

The common 20th century textbook argues that the orbit of a body occurs by a means other than an inertial centrifugal force. In fact, it is simply assumed that a body responds to the inward "centripetal" force and is deflected inward, thus moving along a curved trajectory. However if we examine the case of a massive fly-wheel of the kind used to store mechanical energy in power plants and even now in electric vehicles, we can see that the centripetal force does not provide the complete picture. The conventional view states that if one were attached to the wheel and were observing a piece of it somewhere near the rim, then one would be able to sense that the structure was in tension as if every particle was being pulled outward, possibly expanding the diameter of the fly-wheel. However it would be argued that this outward centrifugal force was an illusion, only observable as a consequence of the fact that you are sitting on the rotating object yourself, which is not considered to be a valid inertial observation frame. According to these same texts, if you are standing still in a valid frame such as the one that contains the axis of the fly-wheel (on the floor in the laboratory), then there is supposedly no outward centrifugal force on any part of the disk. You will however be quite surprised if you then turn up the speed of the motor sufficiently and the wheel explodes. Every rotating object has a critical rotational velocity called its burst speed at which outward inertial forces overcome the chemical bonds that keep the object together. The wheel will be seen to explode from any frame of reference you choose to inhabit and you would be forced to admit that centrifugal forces are both very real and observable from any point of view and are in fact quite dangerous.

How do we know that the centrifugal force is a force of inertia? If we examine the outward force on a particle revolving around a centre, we see that this force which changes direction all the time is clearly not an inward force of gravitation directed toward the centre of mass of a large object. It exists regardless of whether the particle carries an electric charge or is a magnetic atom. Therefore it is neither an electric nor a magnetic force. Excluding nuclear forces, which reside inside atoms and nuclei and do not reach outside the atom, the only other force we know is the force of inertia. Hence by elimination of all other forces, the centrifugal force must be a force of inertia.

Most importantly, it can be shown that the centrifugal force opposes the acceleration relative to the fixed stars. To see this we take note of the fact that circular or elliptic orbiting of a particle around a centre of rotation, requires the continuous deflection of the straight line path of the particle, relative to the fixed stars. This is called radial acceleration to distinguish it from the linear acceleration of objects accelerating in the direction that they are moving. By the principle of dynamic equilibrium, the radial inward acceleration is precisely opposed by a radial outward force of inertia. The centrifugal force that opposes radial acceleration does precisely this.

The conventional textbooks also discuss another apparent effect, called the *Coriolis force*. It is frequently called an inertial force, but in fact is simply an observational illusion. It is named after the eminent French mathematician Gaspard Gustave de Coriolis (1792-1843). While he was primarily involved in the mathematics of engineering including the design of machinery and the development of the concepts of work and kinetic energy, he is best remembered for his deduction of an apparent acceleration, which is often erroneously described as a force. He was concerned that one could not use Newton's laws of dynamics to explain observed motions if one was standing in a rotating reference frame. This is intuitively obvious if you imagine trying to throw a ball to a friend when both of you are riding on a playground merry-go-round. If the platform is spinning, then if you aim directly at your friend, he will not be in the same place when the ball arrives at the desired spot. In general, during the time the ball was in flight, your friend will have moved to the right or left of where you aimed. To actually get a ball to him, you will have to aim a certain amount to one side of his body. The mathematics of how far to aim to the right or left was worked out by Coriolis.

Figure 5.4 : Gaspard Gustave de Coriolis

As a gifted mathematical engineer, Coriolis conjectured a mysterious force, or more accurately an acceleration which depended on the rotational speed of the frame of reference (the merry-go-round) and the velocity of the object (the speed of your throw). So if one knows these two parameters, one can anticipate the deflections that occur when standing in a rotational frame. However, anyone standing still in the playground will see the two people on the merry-go-round who may or may not be successful in their throwing routine, but in all cases the ball will be seen to fly straight relative to all of the stationary objects in the vicinity. Therefore there is no unanimously observed force on the ball during its flight and the Coriolis force truly is not real. Further the amount of apparent deflection does not depend on the mass of the ball, clearly demonstrating that the effect is not caused by a force. The Coriolis acceleration is only ever observed from a frame of reference which is rotating with respect to the fixed stars.

The Coriolis force is therefore purely a mathematical tool of kinematics which allows one to predict how physical effects will look when observed from a rotating frame of reference. It is certainly not the force of inertia. It is often said that it is the Coriolis force that makes clouds spiral in certain directions in weather systems. One must remember that we are observing our weather not from the frame of the fixed stars, but from satellites that are in geo-stationary orbit (stationary in the earth's rotating frame) Under these conditions the Coriolis analysis is an acceptable and useful conjecture for the prediction of cloud movement.

Another well known example of the Coriolis acceleration is the observation of a spinning gyroscope suspended in a gimbal support like the one shown in figure 1.4 in chapter 1. The gimbals ensure that any external forces are equally impressed on both ends of the axis of the flywheel which is held in very low friction bearings. A gyroscope in this configuration sitting on the surface of the earth and watched by someone standing next to it will in general be seen to change the direction of it's axis of rotation gradually throughout the day. It turns out that under these conditions, the axis of the gyroscope is in fact remaining fixed in the sense that it will remain aligned between one group of distant galaxies on one end of its axis and another set at the other end. The fact that the earth is revolving on its axis as well as orbiting the sun has no effect on this alignment which is held steady by the forces of inertia which act between the accelerating atoms of the fly-wheel and the vast amount of isotropically distributed matter in the distant universe. The rotation of the gyroscope axis in the earth bound lab is a manifestation of the so called Coriolis force or acceleration. The value of the acceleration tells the observer nothing about the properties of the gyroscope but does allow him to determine his own rotational motion with respect to the fixed stars. The fact that this apparent axis rotation in the laboratory is independent of the mass or rotational frequency of the gyroscope demonstrates that this motion is not caused by a real force.

An ingenious device which demonstrates the Coriolis acceleration is simply a very long and heavy pendulum which when set moving will always remain in one plane with respect to the fixed stars. Anywhere on earth other than on the equator, the plane of the pendulum is found to rotate slowly throughout the day. One can therefore measure the Coriolis acceleration on it and deduce the rotation of the spinning earth. This was in fact the first measurement of the earth's rotation without observing the sky. It achieved great notoriety when performed publicly by Lèon Foucault in the Panthèon in Paris in front of Emperor Louis-Napolèon in 1851.

The modern gyroscope was also invented by Foucault and represents a much more compact and rugged system than a pendulum and is ideal for navigation during space travel. When Neil Armstrong led the first expedition to land on the moon, he had spinning gyroscopes aboard. If these devices started altering their alignment with respect to his spacecraft, then he could measure the apparent Coriolis acceleration and knew that it was caused by his ship rotating with respect to the fixed stars. He could then fire rockets to compensate and ensure he was travelling in the orientation he desired. Gyroscopes are a wonderful navigational tool and are common in airplanes and ships, but are slowly being replaced by solid state electronic devices that perform the same task.

Only in the 20th century in which the centrifugal force was incorrectly deemed fictitious, did the obviously unreal Coriolis force become discussed alongside it. Since, the confusion over the centrifugal force did not deflect its original status as an inertial force, it seems that the Coriolis force incorrectly became also known as an effect of inertia. This resulted in making the Coriolis force seem more real than it should and the centrifugal force became seen as more fictitious. Fortunately, now that we are in a position to state unequivocally that the centrifugal force is real for all observers and we can ascertain its source, we can clarify the roles of the apparent guises of the force of inertia.

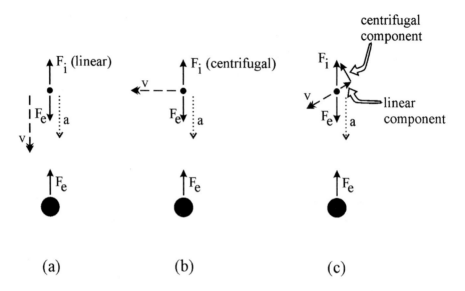

(a)   (b)   (c)

Figure 5.5 : The simultaneous inertial force, F_i, on a particle when acted on by an external force, F_e. The angle of the external force to the particle's velocity vector determines whether the inertial force is called (a) linear, (b) centrifugal or (c) a combination

The real force of inertia can be described in two forms, linear and centrifugal. If an object is moving in the same direction as the force that is accelerating it such as an apple falling straight toward the centre of the earth, then the opposing force is simply linear inertial resistance, as shown in figure 5.5(a). Conversely, if the object is moving in a direction which is perpendicular to the applied force acting on it such as the moon in a stable orbit, then the inertial force, which will also be perpendicular to the motion is called a centrifugal force, figure 5.5(b). In general as shown in figure 5.5(c), an applied force can act at any angle on an object with respect to its velocity vector. Figure5.5(c) depicts an object moving in an inward spiral toward a body with insufficient velocity to sustain a stable orbit. There will still be a single inertial force, F_i , equal and opposite to the applied force, $\mathbf{F_e}$. However for the sake of analysis it is convenient to note that it can be considered to have a component of centrifugal force and a component of inertial resistance with respect to the direction of motion. In all cases, the linear force controls the magnitude of the particle's velocity and the centrifugal force controls the deflection from a straight line trajectory. While these two guises of inertia help to give understanding to the role that inertia plays, it is important to emphasize that they are just mathematical constructions. There is only one force of inertia acting on a body and it is always exactly opposing an externally applied force.

Relative Rotation

Newton assumed, and space travel has confirmed, that a spinning body such as a gyroscope, anywhere in the solar system is subject to centrifugal forces of the magnitude described by his laws. At any time, the axis of revolution of this rotating and otherwise passive body will point from one particular fixed star to another. The axis will not deviate from this direction unless the body is acted upon by external torques. While the axis of the spinning object (gyroscope) points in a given direction relative to the fixed stars, the earth is spinning about its own axis and also orbiting around the sun. Therefore the plane of the radial acceleration vector of an atom in a gyroscope (the plane that contains the atom and is perpendicular to the axis) does not stay fixed relative to the surface of the earth. Not surprisingly, the circular motion of this particular atom can only be treated in the Newtonian dynamics as moving relative to the fixed stars.

In Newton's extrapolation of his new world system, he predicted that centrifugal forces pulling on the spinning earth enlarge the earth's diameter at the equator by 1 part in 230 compared to the polar diameter, which, remarkably, is only 20% greater than the modern measured value. These measurements can be made from space or by earthbound surveying techniques. Faced with such an obvious proof of the work that can be achieved by centrifugal forces, the modern textbooks still deny the existence of the inertia force. For this example their argument runs something like this. All motion is relative motion of one set of bodies relative to another set of bodies. and therefore we can also speak of relative rotation. A bystander in the playground will say the merry-go-round is turning while he is standing still. However, the rider on the merry-go-round may think he is at rest and the world turns around him. What if we take the view that the earth is standing still and the sky of the fixed stars is revolving around it. Then according to the conventional view, no centrifugal forces act on the earth and its equatorial diameter should be equal to the polar diameter. This would mean that the world is not governed by unalterable laws of physics, but by our minds which can arbitrarily decide whether the earth is rotating or whether the universe is spinning around it.

There has to be something wrong with this particular relativity argument. The rider on the merry-go-round feels the centrifugal forces which try to throw him off, never mind how hard he thinks that it is the rest of the universe that turns and he is stationary. Similarly, the equatorial diameter of the earth did not decrease simply because a philosopher named Aristotle claimed that the earth stands still in the center of the cosmos.

To placate the fallacy, the principle of relativity should be expressed as;

The force laws of nature remain the same regardless of which particles of the universe are considered to be at rest.

We have proposed that all forces including the force of inertia are interactions between particle pairs. In that case, it is easy to see that the force of inertia, treated as an interaction of a particle on earth with all the particles in the distant universe, is the same for the earth spinning inside the sky of the fixed stars, or the sky revolving around the earth.

On the other hand if, with Newton and his inertia force caused by absolute space, we believe that the force of inertia is not a matter interaction force, then we are driven toward the idea of a fictitious force which magically disappears when we mentally reduce the earth to absolute rest. The Machian interaction theory of inertia which we develop in the final chapter, does not ever allow the force of inertia to do such a disappearing trick. The reason is that, by Newton's 3rd law, the interaction force between a particle on earth and another in the distant universe is the same whichever of the two is considered to be at rest. Therefore in the Newton-Mach paradigm that we are proposing, there is no artificial distinction between linear and rotational motion.

The Newtonian Legacy

While the post-Einstein modern physics makes a great point of claiming that Newton has been disproved, it is remarkable that engineers and physicists use his laws of dynamics successfully every day. This is not the place to debate the validity of Einstein's so called "corrections" to the Newtonian paradigm for they are based on absolutely minute effects which are irrelevant to every day life and cannot be tested without making sweeping assumptions about the nature of light, space and time. However, it is fair to say that Newton's ideas certainly did need some adjustment to become a conceptually self-consistent paradigm.

As we will see in the next few chapters, philosophers, mathematicians and physicists such as Jean D'Alembert, Immanuel Kant, Joseph Lagrange and Ernst Mach all made slight modifications to the Newtonian laws which made them more of a complete and universal system. Our proposed reworded versions of some of these laws, presented in this chapter, hopefully bring them further up to date with modern knowledge and terminology. However, without an expression for the actual force of inertia acting as a mutual force between every pair of particles that are in relative acceleration, the system remains incomplete. Mach proposed that such a law would hopefully come to light. In the final chapter of this book, we propose a law modelled on Newtonian gravitation which fulfils Mach's requirement. [5.13]

It is almost impossible to express the sheer magnitude and practicality of Newton's discoveries in the field of dynamics. While his study of optics and alchemy may have occupied more of his academic life, they pale into insignificance when compared to the masterful *Principia*. That such a complete treatise could be prepared in a few years, and was immediately recognized by many of his peers as a fundamental breakthrough in understanding is why Newton is revered as possibly the greatest scientist of all time. His laws are still used to build bridges and guide space ships with pin-point precision. We should be very wary of casting unwarranted doubt on this magnificent body of knowledge without very strong proof. It seems more prudent to build upon what appears to be an infallible foundation.

Chapter 5 References

[5.1]   I. Newton, *Principia (1686)*, F. Cajori (Ed.). Berkeley: University of California Press, 1962.

[5.2]   R. S. Westfall, *The Life of Isaac Newton*. Cambridge: Cambridge University Press, 1993.

[5.3]   A. P. French, *Newtonian Mechanics*, 2nd ed. New York: W.W. Norton, 1971.

[5.4]   P. Gondhalekar, *The Grip of Gravity*. Cambridge: Cambridge University Press, 2001.

[5.5]   J. Michell, *A Treatise on Artificial Magnets*. Cambridge: University of Cambridge, 1750.

[5.6]   C. A. Coulomb, *Mémoirs*. Paris: Académie Royale des Science, 1785-89.

[5.7]   A. M. Ampère, *Théorie mathématique des phénomènes électrodynamiques uniquement déduite de l'expérience*. Paris: Blanchard, 1958.

[5.8]   N. C. Barford, *Mechanics*. New York: Wiley, 1972.

[5.9]   P. Graneau, N. Graneau, *Newton versus Einstein*. New York: Carlton Press, 1993.

[5.10]   W. Stukeley, *Memoires of Sir Isaac Newton's Life*. London: Taylor & Francis, 1936.

[5.11]   K. A. Baird, "Some influences on the young Isaac Newton," *Notes and Records of the Royal Society of London*, vol. 41, p. 169, 1987.

[5.12]   J. B. Barbour, *Absolute or Relative Motion?* Cambridge: Cambridge University Press, 1989.

[5.13]   P. Graneau, N. Graneau, "Machian Inertia and the Isotropic Universe," *General Relativity & Gravitation*, vol. 35(5), p. 751-770, 2003.

Chapter 6

A Century of Consolidation

The Early Practitioners of Newtonian Dynamics

The seventeenth century was the heroic century of science during which quantitative experimental science was born and two thousand years of Aristotelian physics teaching was overthrown. Even though it commenced with the burning at the stake of the heretic Giordano Bruno for proclaiming radical opinions which included the Copernican world model, it was on the whole not a bloody transition. This contrasted with the extremely violent religious and political wars which were occurring throughout Europe which more directly moulded the fortunes of people, nations, and empires. Some of the ingenious scientists of the time were left to pursue their discoveries in peace, like Newton, while the brave Galileo stubbornly faced the Inquisition and accepted imprisonment for his views. Most others trod a very careful path which allowed them to express their views but only in terms that would not upset the religious hierarchy. Fortunately while the dominant stage on which human drama was acted out remained the conquest of territory, the destruction of old governments, and the building of new ones, the calmer activities of science had a more lasting effect on the well-being of the human race. However if the seventeenth century provided the birth of experimental and quantitative science, it was in the eighteenth that these ideas were consolidated and made practical.

Our lifestyle today is largely the result of the scientific revolution of the seventeenth century which burst open the floodgates of technology and accelerated our civilization. Early in that century the Copernican view of the world gained ground among free thinkers. The earth was no longer the undisputed center of all that mattered and the human race was apparently not the purpose of the existence of the universe. However, Kepler had sailed a precarious course between the highly profitable but subjective field of astrology and the more objective science of astronomy. He had been forced to employ his considerable diplomatic skills to gain access to the data he needed to discover his mathematical formulae which described the motions of the planets around the sun. Shrouding his discoveries in mathematics that were beyond the understanding of most people, including the church leaders, Kepler's achievements were not ignored but his career was not a march of glory, as it should have been. He seemed to be constantly dogged by fear and in the end was lucky to be able to save his mother from the burning pyre.

Galileo, despite his scientific triumphs and huge popular acclaim, was tried, sentenced, and forced to live out the last years of his life in house arrest. During this repressive episode, the brilliant Descartes elevated rationalism above the authority of scripture and, at the same time, dreamt up a cosmos based on a structure that lay beyond our senses. Nevertheless, he also spent most of his life on the run from those who tried to stem the revolution. In the latter part of the century, Newton laboured away in monastic seclusion between flashes of exhilarating insight and deep doubts and depression. The *Principia* gained him immediate fame, but he was haunted to the end of his days by the charge of promoting *occult* attractions and atheism. Nevertheless, an age of enlightenment was dawning and the eighteenth century was a well deserved period of unfettered development of the new physics paradigm.

The eventual institutional acceptance of Newton's action at a distance was in no insignificant way due to the efforts of the eloquent French poet, playwright, and master of the written word: Françoise Marie Arouet (1694-1778). He had attained notoriety when, only twenty-three years old, he served a one year prison term for having ridiculed royalty. In the Bastille he created his famous pen name, Voltaire. His troubles with the ruling class continued after his release from prison. In 1726 he was exiled to England and found himself in London at the time that Newton was buried in Westminster Abbey like a king. It inspired the ever curious Voltaire to study Newton's philosophy and physics. and to understand what had led to such glorification of a scientist.

Figure 6.1 : Voltaire

Back in France, following a further indiscretion with the aristocracy, he retreated with his mistress and colleague, Mme. Emilie Du Châtelet to her husband's Chateau de Cirey in Alsace where he could easily slip across the German border if pursued by the authorities. There, they performed experiments together in her laboratory and wrote a book [6.2] for the general public about Newton's great achievements. Besterman [6.1] described the success of this publication *Elemens de la philosophie de Newton* [6.2] by quoting a Jesuit priest: "All Paris resounds with Newton, all Paris stammers Newton, all Paris studies and learns Newton."

For his efforts, Voltaire was elected a Fellow of the (British) Royal Society. In a letter of thanks to the President of the Society he wrote of his dismay regarding his unenlightened French colleagues [6.1]:

"But the liberty of the press was fully granted to all the witty gentlemen who teach'd us that attraction is a chimera, and vortices are demonstrated ..."

The "witty gentlemen" were the French professors of natural philosophy who continued to adhere to Descartes' ether whirlpools and taught that simultaneous gravitational attraction between separated objects was a fabrication of the mind. The following comment was found in Voltaire's notebook [6.3]: "Before Kepler all men were blind. Kepler had one eye. Newton had two."

Figure 6.2 : Mme. Emilie Du Châtelet

Voltaire was clearly very impressed with the tolerance and freedom of speech that he observed in England. His greatest service to society may have been his efforts to introduce this attitude to France and indeed the rest of Europe. He may have done more than anyone to publicise the Newtonian revolution, but there were even more important proponents who performed the highly dedicated tasks of codifying Newton's laws and making them more practical and mathematical. There developed a movement to find generalised equations that would allow the solution of any mechanical problem. This contrasted with the earlier approach in which each situation was resolved by a unique geometrical construction. As the applications became more complicated than astronomy and began to include the fields of hydrodynamics and vibrations, the geometrical approach became too unwieldy and dynamics became the stomping ground for some of the finest mathematicians of the modern age.

The first giant to take great strides on this mission was the formidable mathematician, Leonhard Euler (1707-1783). Born and raised in Switzerland, he nevertheless spent most of his professional life in the St. Petersburg Academy of Science, which had just been founded by Catherine I, the wife of Peter the Great. Euler is the most prolific writer of mathematics that has ever existed. He developed many of the symbols and concepts that mathematicians now use every day. Most importantly for the further development of the science of mechanics, he integrated Newton's fluxions and Leibniz's calculus into a more manageable tool that is the version of calculus that is taught today which greatly facilitated the analysis of real problems. In his writings on forces and motion, he discussed the *vis inertiae*, saying [6.4]:

"The force of inertia is the force that exists in every body by means of which that body persists in its state of rest or of uniform motion in a straight line. It should therefore be reckoned by the force or power that is necessary to take the body out of its state. Now different bodies are taken out of their state to similar extents by powers which are proportional to the quantities of matter that they contain. Therefore their forces of inertia are proportional to these powers, and consequently, to the quantities of matter."

While Newton had ascertained that the force of inertia was caused by a change in momentum, or motion as he called it, Euler seems to have expanded on this idea. He was developing a framework that makes it clear that the mass, which he called the quantity of matter, controls how big a force must be to achieve a certain acceleration. His use of the word power demonstrates that there still was not an accepted understanding of the meaning of force. Euler merged the concept of what we now call a force such as gravity or collision with a notion that was then discussed as *living force*, which can best be described in modern terms as momentum. Nevertheless, he did not hesitate to state that the inertial forces are a direct consequence of applied forces and are of the same magnitude. He seemed to assume that it is obvious and does not need spelling out that the two forces are in opposition.

The most famous exposition of the model specifying counteracting applied and inertial forces was produced by Jean La Ronde d'Alembert (1717-1783), the French mathematician and philosopher. Despite being abandoned as a baby and raised by a humble family, this did not prevent him from receiving a solid mathematical education. He made significant progress in the field of dynamics in the wake of Newton's *Principia* and clearly saw his role as a major player in the consolidation of these new ideas. He is quoted as writing [6.5]

"Once the foundations of a revolution are laid, it is almost always in the following generation that the revolution is completed: rarely earlier, because the obstacles disappear of their own accord rather than give way; rarely later, because once over the barriers, the human spirit often advances faster than it itself wishes, until it reaches a new obstacle which forces it to rest for a long time."

D'Alembert's most important work was published in 1742 under the title *Traite de dynamique*. It comprised extensive analysis of forces and motions. He famously used Newton's force laws to perfect the calculation of the precession of the equinoxes. This occurs because the earth is not spherical, but is larger around the equator than a great circle connecting the poles. This bulge causes gyroscopic behaviour of the spinning earth relative to the sun and the moon. The effect causes the axis of the rotation of the earth to change direction so that it points to different directions in the frame of the distant fixed stars. At the moment, the northern end of the axis points to what we now call the Pole

Star or North Star, but this was not the case several thousand years ago. The axis describes a cyclical precession which will return to the same position every 26,000 years. Remarkably, this subtle motion has been known and accurately recorded since the time of the ancient Greeks. Newton had already started to solve this problem in the *Principia*, but appears to have fudged his calculation His estimate of the lunar mass was inaccurate and therefore he incorrectly calculated the relative contributions from the sun and the moon, but yet he mysteriously came up with a figure that precisely matched the observed value.

Figure 6.3 : Jean La Ronde d'Alembert

It is frequently claimed that d'Alembert's greatest achievement was a principle which he formulated and which is known by his name. In modern dynamics, which was initiated by Mach, this principle is called the condition of dynamic equilibrium. D'Alembert's principle is really nothing more than a restatement of Newton's second law combined with the latter's assertion that the force of inertia on a body is equal and opposite to the sum of external forces which cause the acceleration of the body relative to the fixed stars. Since the sum of the external force and the opposing inertia force always comes to zero, dynamic equilibrium is ensured.

It is not at all clear at first why this restatement required the special moniker of d'Alembert's principle. For the simple issues involving two bodies such as a satellite orbiting the sun, his principle provided no further understanding. However, d'Alembert's goal was to allow the application of Newtonian physics to a broader range of phenomena, namely contact forces and the interactions of large numbers of particles which required more subtlety than Newton had previously applied.

There was general interest in how a force applied to a macroscopic body at one point can move the entire object as a whole such as a block of wood being pulled by a thread. D'Alembert formally treated a large body as being composed of many smaller particles which was still an unusual concept in the pre-atomic age. Under conditions in which the body remains intact, the microscopic particles are in a state of relative static equilibrium since they do not accelerate with respect to each other, whatever the overall motion of the combined lump. In order to understand how this compound body could become accelerated, d'Alembert had to re-emphasize Newton's definition of inertial force to explain that as a particle was pulled or pushed by a neighbouring particle, it resisted the acceleration. Only in this manner can the entire mass of a body affect its overall acceleration even when an external force acts at a single point. The understanding of how particles remain fixed relative to each other even when the whole object is accelerating with respect to the frame of the fixed stars led to d'Alembert's principle becoming a breakthrough in the science of dynamics.

Following an account given by Dugas [6.4], d'Alembert treated the force that accelerates an object as an "obscure and metaphysical" concept, thus explaining why he sought to eliminate it by concentrating on the notions of equilibrium and motion. Modern physics, however has demonstrated the utility of the concept of force and thus d'Alembert's principle appears somewhat redundant. His mathematical derivation of the principle was very complicated and his aversion to force as the cause of motion somehow concealed his subsequent use of the concept of the force of inertia. Only later did the essence of the principle emerge. In 1959 Meriam [6.6] wrote:

"Thus, if a fictitious force equal to *ma* (mass times acceleration) were applied to the accelerating particle in the direction opposite to the acceleration, the particle could then be considered to be in equilibrium under the action of the real forces *F* and the fictitious force (*ma*). This fictitious force is often called an inertia force, and the artificial state of equilibrium is known as *dynamic equilibrium.*"

Except for the word *fictitious*, this is a clear formulation of d'Alembert's principle and, indeed, Newton's definition of the force of inertia opposing acceleration.

That this force should be called fictitious in modern textbooks is absurd because, in the guise of the centrifugal force, it is actually responsible for the stability of the universe. Moreover, the force of inertia can do real work. Consider a weight resting on the flat bed of a truck traveling at constant speed along a horizontal road. When the driver applies the brakes, the weight slides forward propelled by Newton's force of inertia. In overcoming the friction between weight and floor, this so-called fictitious force generates heat. It is left to the readers to question whether they believe that real heat can be generated by a fictitious force.

A similarly misleading form of d'Alembert's principle was given by Synge and Griffith [6.7]:

"The reversed effective forces and the real forces together give statical equilibrium."

In this quotation, the word *effective* takes the place of *fictitious*. To call the force of inertia "effective" is an improvement, but still not sufficient to describe its very real existence.

Many other writers starting with some of his contemporaries have also described d'Alembert's principle as resulting in the static equilibrium of forces acting within a body or other closed system. D'Alembert never accepted this criticism for he argued that all motions involving the force of inertia are accelerations and are thus by definition not static situations.

At least the Synge and Griffith restatement of d'Alembert's principle admits that the force of inertia has equal standing with forces like gravitational attraction. Therefore, in the same way that Newton demonstrated how the force of inertia controlled the non-local force of gravity and thus was responsible for the stability of the solar system, d'Alembert similarly explained how inertia modulated the acceleration of solid compound objects under the influence of local contact forces.

Newtonian physics received strong support through the writings of the German philosopher Immanuel Kant (1724-1804). He was a popular teacher and highly respected thinker of the eighteenth century who hardly ever left his native city of Königsberg in East Prussia. He was a true early professor of philosophy in the sense that, as well as espousing his own reasoning, he had immense knowledge regarding the human history of philosophical thought and had the gift to convey the progression of human thought to his students. His method of reasoning was highly novel but largely non-controversial. He was a fully accepted member of the academic establishment, but late in life he quarreled with the Prussian state about religious matters and suffered censure as a consequence.

In one of his early books [6.8], *Universal natural history and the theory of the heavens,* he said: "Give me matter and I will build a world." No ether, no non-material particles, no free energy flying through space were required. This appears to have been self-evident to the Newtonians who laid down a system of physics in which only the laws of attraction and repulsion drove all changes in the material world.

To Kant inertia was not so much a force of resistance to change of motion, but a behavioral property of matter. In his *Metaphysical foundations of natural science* [6.9], he wrote:

"The inertia of matter is and signifies nothing else but the lifelessness of matter in itself. Life means the capacitance of a substance to act on itself, from an inner principle, of a finite

substance to act on itself, and of a material substance to determine itself to motion or rest as alterations of its state. ....
Therefore all matter is lifeless. This, and nothing more, is what the proposition of inertia says."

Figure 6.4 : Immanuel Kant

Just like Newton's first law of motion, this description of motion unaffected by external forces does not preclude nor support the existence of forces of inertia. Nevertheless, it is often still described as the principle of inertia. Kant's concerns demonstrate that it is by no means easy to explain why animals can propel themselves, while lifeless bodies are condemned to rest, unless acted upon by external forces. On closer examination one finds that the living body has the capability of creating internal muscular stresses by burning fuel. It can therefore set up forces of attraction and repulsion which expand or contract the body. As the knee straightens when we walk, the muscles make the leg longer, propelling the body forward with the help of friction under foot. It brings an external force into play, that of friction, which is absolutely essential for moving the center of mass of the body.

For this reason, an astronaut in empty space is incapable of moving himself however he flails his arms and legs. To accelerate toward or away from his spaceship, he must have a physical connection or else jettison material, usually from a jet pack. The stresses in a living body are composed of many equal and opposite inter-particle forces which, as far as the body as a whole is concerned, add up to zero and cannot move its centre of mass. The forward acceleration caused by the frictional interaction force with the ground is controlled by the force of inertia without which the body would simply push off with infinite acceleration.

It is not only living bodies that are capable of moving themselves, lifeless bodies can do this too if internally locked up stresses are suddenly released. Take a rubber ball pressed down by hand to the surface of the earth. When the hand, which sets up the internal stress, is quickly taken off, the *dead* ball will jump in the air. It could not do so without the external reaction force supplied by the earth. Kant was mistaken when he claimed that self-propulsion depended on life, but it was not so obvious in an age before the advent of hi-tech materials and machinery.

Curiously Kant, who did not understand the nature of the force of inertia, nevertheless made major positive contributions to Newton's science in the form of his unusual ideas about space and time. Ultimately he took up a position half way between Newton and Mach. He denied that absolute space and time are physical realities, as Newton had erroneously assumed. At the same time Kant was not ready to deny space and time a place in physics. In his *Critique of pure reason* [6.10] he wrote:

"Space is not an empirical concept which has been derived from outer experiences. For in order that certain sensations be referred to something outside me, and similarly in order that I may be able to represent them as outside and alongside one another, and accordingly as not only different but as in different places, the representation of space must be presupposed."

We cannot see, hear, feel, smell, or taste space. It lies completely outside our sphere of sense perception. This means it is not something we become acquainted with through experience. A child cannot be taught what space is. Nevertheless, every human being seems to be familiar with the concept of space and can visualize objects in space. Therefore the concept of space must exist in our mind *a priori*, as Kant said. The mind presupposes it. Without this *a priori* concept of space in our brain, we would not be able to orient ourselves. The sense of orientation is essential to animal life on earth and is probably the reason why evolution has reinforced the *a priori* notion of space in our brain.

Distance is not the same thing as space. It is a human invention which can be measured by counting subdivisions on a ruler which has been deemed to represent a standard length. There is meaning in the statement that one object is twice as far away as another, but nobody can ever define what a metre really represents. Distance is purely the relative position of bodies and particles and therefore has nothing to do with Newton's absolute space. Kant pointed out that while we can imagine empty space devoid of all matter we cannot think of the absence of space! *A priori* knowledge therefore remains permanent knowledge.

Kant was adamant in believing that space did not represent a property of anything. Extension of a material object was measured by defining distances between points on the object. These were relative measures grasped by the senses and they had nothing in common with the abstract concept of space. Precisely this philosophy was to clash later with Einstein's idea of curved space which had material properties and guided material objects. According to Kant, space conveys nothing about the spatial relationship between bodies, which are purely defined by distances.

Newton considered time to flow incessantly, like a river, at constant speed. To him time was independent of everything else. Since the notion of *flow* requires *a priori* understanding of time, Newton's definition of absolute time was circular and imperfect. The Kantian view was [6.10]:

"Time is not an empirical concept that has been derived from any experience. For neither coexistence nor succession would ever come within our perception, if the representation of time were not presupposed as underlying them *a priori*. .... We cannot, in respect of appearances in general, remove time itself, though we can quite well think time as void of appearance. Time is, therefore, given *a priori*."

If we look at the measurement of time, we find what is measured are the relative periods of repeating events such as intervals between sunsets, swings of a pendulum, oscillations of an atom, and so on. *Period* is as much a relative measure of time as *distance* is a relative measurement of space. There is no time to be observed except the relative periods between matter events. Time itself is not accessible to our senses and by the same logic must exist as *a priori* knowledge. In this way Kant admonished the *absolutes* of Newtonian physics and paved the way to the principle of relational mechanics which reached its zenith with the development of Mach's principle which is the subject of the next chapter.

Again one might ask why should nature have chosen to put this abstract concept of time into the mind of man before birth? The answer could once more be necessity. It is not enough for us to know where we are in relation to our abode, we must also have an idea of how long it may take to get back. Also in order to comprehend motion of our bodies and things around us, we need to instinctively understand the notion of speed which by definition requires simultaneous appreciation of both distance and time period. *A priori* space and time are likely to be the essential tools of our sense of orientation which allows us to crudely find our way even without a compass, measuring stick or watch.

From Kant's philosophy one may conclude that space and time are something one needs for the business of living. They do not represent physics in the traditional sense of being observable properties of our material world and are often confused with the relative measures of distance between particles and repetitive event periods. Both of the latter are very much a part of physics. However the Kantian developments made these issues clearer and allowed Newtonian mechanics to free itself of its weaknesses involving absolute space and time.

After the publication of the *Principia*, mathematics became an ever more important aspect of physics for calculating the future motions of celestial bodies and objects on earth. These issues were not only of academic interest, but had great economic significance since they directly affected both the science of ship navigation as well providing a theoretical framework for the machinery that was to drive the upcoming industrial revolution. With such huge benefits beckoning, a formidable collection of analytical techniques was developed during the eighteenth century. Of this new group of pioneers, the names of two French mathematicians stand out. They applied the new methods to celestial mechanics involving more than two bodies and the consolidation of Newton's cosmology in general. In addition, they tackled more earth bound mechanical issues which were to become the bedrock of modern engineering.

One of these scientists was Joseph Louis Lagrange (1736-1813). At the age of nineteen, he became a professor of geometry at a military academy in his native Italy. His aptitude was quickly recognized by the distinguished mathematicians in Berlin and Paris. d'Alembert was very impressed with Lagrange's celestial calculations and went to great efforts to secure a post for him in Berlin, working under Euler. However, both Lagrange and Euler were at the time developing the elements of calculus and differential equations at such a prolific rate, that there seems to have been a personality clash that precluded their close collaboration. Lagrange eventually succeeded Euler in Berlin when the latter returned to St. Petersburg in 1766. In 1787, Lagrange moved to Paris to become a member of the Académie des Sciences and a year later his most important work, *Analytical Mechanics*, was published, in which he summarized all of the work on mechanics that had occurred since Newton. It was a remarkable achievement in that the book contained no diagrams and the entire science was described in mathematical equations, emphasizing the first real exposition of differential equations. In 1793, the dark forces of the French Revolution swept through the Académie. Much of the institute was closed down with many of the great minds sentenced and executed. As a mark of how highly Lagrange's skills were valued, he was reprieved and made president of a commission which introduced the metric system of measurement. Emperor Napolèon-Bonaparte greatly admired him and upon his death he was buried with great honour in the Pantheon.

Pierre Simone Laplace (1749-1827) was a contemporary of Lagrange and a protégé of d'Alembert and no less famous a mathematician. He made a thorough study of gravitation and investigated the stability of the solar system. Newton had come to the conclusion that the stable orbits of the planets would deteriorate with time due to the interaction of the planets with each other. This led him to believe that God would have to intervene occasionally to correct the orbits. The calculations made by Laplace indicated, however, that most of the changes were cyclical, and over long periods of time the solar system would remain unchanged. These calculations ignored any decrease in the mass of the sun.

In 1799, Napolèon-Bonaparte appointed Laplace to be his minister of the interior. He was in his post for only six weeks because the mathematician administered the affairs of state with differential equations and Napolèon objected to the "spirit of the infinitely small" being brought into government. Laplace nevertheless remained in favour with the Emperor and was made a marquis.

The two French mathematicians, Lagrange and Laplace, made good use of Newton's force of inertia without which the solar system would collapse. However, they did not in any way advance the understanding of the nature or cause of inertia. For them, Kant's space and time considerations came too late to have influenced their thinking and acceleration relative to absolute space was still believed to be the cause of the inertia force. They nevertheless provided the mathematical foundation upon which Mach built to make his radical and bold assertions in the next century.

Chapter 6 References

[6.1]   T. Besterman, *Voltaire*. New York: Harcourt, Brace & World, 1969.

[6.2]   Voltaire, *Elemens de la philosophie de Newton*. Amsterdam: Etienne Lechet, 1738.

[6.3]   C. A. Ronan, *Galileo*. New York: G.P. Putnam's, 1974.

[6.4]   R. Dugas, *A History of Mechanics*. New York: Dover, 1988.

[6.5]   I. B. Cohen, *Revolution in Science*. Cambridge, MA: Harvard University Press, 1985.

[6.6]   J. L. Meriam, *Mechanics Part II: Dynamics*. New York: Wiley, 1959.

[6.7]   J. L. Synge, B. A. Griffith, *Principles of Mechanics*. New York: McGraw-Hill, 1949.

[6.8]   I. Kant, *Universal Natural History and the Theory of the Heavens*. Ann Arbor, MI: University of Michigan Press, 1969.

[6.9]   I. Kant, *Metaphysical Foundations of Natural Science*. Indianapolis, IN: Bobbs-Merrill, 1970.

[6.10] I. Kant, *Critique of pure reason*. London: Macmillan, 1963.

Chapter 7

Mach's Magic Principle

The Unique Inertial System

Ernst Mach (1838-1916) was a highly respected and successful academic of the Austro-Hungarian empire. He was a professor of physics in Graz (1864-67) and Prague (1867-95) after which he was appointed professor of inductive philosophy at the University of Vienna. He held this position until 1901 when he was elevated to the peerage of the Empire. He was one of the eminent members of the positivist movement which held that no statement of natural science is tenable unless verified empirically. To him, empirical knowledge was that which the human mind acquired through sense perceptions. He was rigidly opposed to the notions of metaphysical and *a priori* knowledge and on these grounds, flatly rejected Newton's absolute space and time.

Mach's attitude was reminiscent of Galileo who had tried to make experiment the ultimate arbiter of scientific pronouncements. He was probably uneasy with the abstract concept of electromagnetic field theory, which was being formulated during his lifetime, and which Einstein completed early in the twentieth century. Mach said very little about fields, for he considered electromagnetism to lie outside his expertise. While he wrote books about almost all other aspects of physics, he studiously avoided the surging science of electricity.

In his youth, Einstein claimed that Mach had provided the initial inspiration for the special theory of relativity. It is surprising therefore that the name of Ernst Mach is a now a thorn in the flesh of modern physics. It is not found in the indexes of such famous works as Bertrand Russell's *History of modern philosophy* [7.1], the *Feynman lectures of physics* [7.2], and Hawking's *A brief history of time* [7.3]. By 1922, Einstein was breaking away from Mach, the idol of his youth. In a lecture delivered in Paris, the creator of the relativity theories said [7.4]:

> "Mach's system [consists of] the study of relations which exist between experimental data; according to Mach, science is the totality of these relations. That is a bad point of view; in effect, what Mach made was a catalog and not a system. Mach was as good at mechanics as he was wretched at philosophy"

There is no room for Mach's views on inertia in Einstein's general theory of relativity. This was the primary reason why Mach's principle has had to be purged from modern physics. A parallel case arose in the special theory of relativity where Ampère's celebrated electrodynamic force law has been written out of textbooks, although it is in full agreement with all relevant experimental facts and explains many experiments which relativity theory can not handle.

Figure 7.1 : Ernst Mach

Ironically, it was Einstein who first mentioned the phrase *Mach's principle*. Mach never formulated a precise statement of his principle, but the idea emerged gradually through his critical discussion of Newtonian dynamics. Had he formulated the principle of which Einstein spoke, the cautious Mach might have said:

Mach's Principle

The inertia force on particles and bodies on earth and in the solar system is due to their acceleration relative to all matter residing outside the solar system.

Mach often referred to this distant matter as the *fixed stars*. He was of course aware of the possibility that additional matter existed beyond the visible universe as known at his time. The additional matter would have been expected to consist of yet more stars. Since there was already no detectable relative motion between the easily visible stars, there was no reason to believe that the positions of the more distant invisible stars were not also fixed relative to the visible ones. So the term *fixed stars* was at Mach's time a justifiable collective reference to all matter outside the solar system, it being understood that the sun was one of the fixed stars.

To all intents and purposes, therefore, the fixed stars formed a rigid system of parcels of matter. If a particle on earth was accelerating or decelerating with respect to one fixed star, its relative acceleration to all the others was immediately determined because of the perceived rigidity of the fixed star system. So if a particle accelerated toward a particular star, it would accelerate away from a star lying in the opposite direction. The fixed star reference system was not far removed from the concept of absolute space, as Newton himself admitted, but it was fundamentally more philosophically acceptable because it was based on matter. So why not use the fixed star reference frame for inertia calculations, rather than the intangible concept of absolute space? This was Mach's point, and with it he created the unique inertial system of the *fixed stars*.

The first spiral galaxy was discovered in 1845, but not until 1927 was it known with certainty that the Milky Way was also a spiral galaxy. Mach probably did not appreciate that the fixed stars with which he was familiar were primarily contained in a disk shaped envelope. This shape makes the Milky Way an unlikely source of the inertia forces on earth. In our experience, the inertia forces are of exactly the same magnitude in all directions. The isotropy (direction insensitivity) of the force of inertia was the great mystery which propelled Newton toward belief in the physical attributes of absolute space. This property has remained the foundation of all subsequent research into the phenomenon of inertia

Mach recognized that in many instances the acceleration which determined the force of inertia could be referred to the earth. He actually said [7.5]:

"I have remained to the present time the only one who insists upon referring the law of inertia to the earth, and in the case of motions of great spatial and temporal extent, to the fixed stars."

This muddied the waters to some extent. Why replace the fixed stars by the earth? There was of course a good reason. It was much easier to measure the acceleration of an object relative to the surface of the earth, or an earth-bound laboratory. Besides, it was a fact of experience that the acceleration relative to the earth would give the correct inertia force, at least to a good degree of approximation. This occurs because the acceleration of a terrestrial laboratory relative to the fixed stars, due to the earth's motion, is small compared to the acceleration of typical objects which are submitted to inertia force measurements.

A laboratory on the surface of the earth accelerates relative to the fixed stars on account of (1) the spinning of the earth, and (2) the orbiting of the earth around the sun. Both accelerations can be calculated from figures in a standard astronomy textbook such as [7.6] and come to (1) 0.034 m/s^2 and (2) 0.006 m/s^2. The standard unit of acceleration is one meter per second per second. A car accelerating at this rate from standstill would come to a speed of 3.6 km per hour or 2.25 m.p.h. in one second. A body of 100 kg mass on a seat of this car would be pressed with a force of 100 newtons against the backrest of the seat. This is the force of inertia given by Newton's second law of motion with the earth as the reference body. Because of the spin and orbit of the earth this requires a correction of at most 4 newtons, or 4 percent, when calculated with respect to the fixed stars (the sun). So we see that, to a reasonable first approximation, the earth can be used as the inertial reference body, instead of the fixed stars.

Unfortunately we then run into a problem with momentum conservation. Momentum is the product of mass times velocity. The mass is that of the body which acquires the momentum. Velocity has no meaning unless it is expressed relative to a reference body. As discussed in chapter 1, for the sake of momentum conservation, it becomes essential to only consider the velocity relative to the fixed stars. This can be shown with the example of the freely falling apple. As the apple breaks away from the tree and accelerates to the ground at 9.81 m/s^2, it acquires momentum. The inertia force (mass times acceleration) controls the fall of the apple. It would be approximately the same regardless of whether the acceleration is expressed relative to the earth or the fixed stars.

However to obey the conservation requirement, the total momentum of matter in the universe must remain constant, and thus the fall of the apple must be accompanied by the creation of opposite momentum somewhere else. Newton's third law couples the apple to the earth via gravitation. Hence the opposite momentum should be produced by the gravitational pull on the earth toward the apple. However the velocity of the earth relative to itself is zero and so would be its earth related momentum. Therefore with the earth as reference body the total momentum of matter in the universe would change while the apple was falling, violating momentum conservation which is one of the most trusted laws of nature. The only correct relative velocity to be used for momentum conservation is that relative to the fixed stars. The apple momentum is then, at every instant, automatically balanced against the earth momentum thanks to Newton's third law and his law of universal gravitation. To Mach, this was the primary clue that the fixed stars represent more than just a picturesque backdrop to our local environment, but actually play an active role in the physics that we observe on earth.

His great leap was to hypothesize that this same framework of stars was also responsible for causing the force of inertia. We now know that the fixed stars of the Milky Way cannot be the dominant cause of the force of inertia on earth and in the solar system because of their uneven disk shaped distribution around the earth and the sun. They would make the inertia force on an object stronger in some directions than in others which is certainly not the case. So in order to uphold Mach's principle, we must look for huge amounts of additional matter in the universe which is isotropically distributed and at rest relative to the fixed stars that we can see. Using their most powerful telescopes, astronomers have as yet not discovered homogeneously distributed visible stars and galaxies, but it is too early to rule out a suitable matter distribution in the furthest reaches of the universe. Invisible dark matter can also be considered as a possible source of this vast and essential mass distribution.

All of this leads to the conclusion that our home galaxy is not really the unique inertia system that Mach hoped it would be. On the other hand, the fixed stars of our galaxy appear to be virtually stationary relative to the much larger amount of cosmic matter which is the main cause of inertia forces. If so, and since the fixed stars are easily visible, they form a far more convenient reference frame than something that resides at the edge of the visible universe or beyond. In effect the Machian reference frame is therefore visible to us every clear star lit night.

The spiral arms of the Milky Way are known to rotate about the galactic center. This means that all the fixed stars in our galaxy are really accelerating relative to each other. They are continuously changing direction and this will induce centrifugal forces. A true inertial system of particles should be free of relative accelerations. However as far as we know, using figures from [7.6], the centrifugal acceleration of the earth in its orbit around the sun is approximately thirty million times greater than that of the sun around the galactic center. It is on this basis that we feel justified to ignore the relative acceleration between the fixed stars. We now have astronomical evidence that whole galaxies move relative to each other, but apparently we may also neglect inter-galactic accelerations in the computation of inertia forces.

The question which then arises is: does uniformly, or isotropically, distributed remote matter move and accelerate relative to Mach's fixed star system? With all the uncertainties surrounding this remote matter distribution, it seems a pointless exercise to try and predict its motion.

The only reasonable interim conclusion has to be that, since the acceleration of all visible matter relative to the sun is negligible, then in inertia force calculations, we are justified to expect this state of affairs not to be upset by the discovery of additional remote matter.

It is a fact that the strength of the force of inertia, observed in any terrestrial laboratory, is independent of the direction of the external force which causes the matter to accelerate, This is a more compelling argument for the existence of isotropically distributed matter in the remote universe than the sum of astronomical observations made of the outer reaches of the cosmos up to the present time. This does not mean however that astronomical data should be ignored.

For much of the last century astronomers have gathered indirect evidence which points to the existence of large amounts of dark matter, invisible in telescopes, intermingled with galaxies and clusters of galaxies. The dark matter would have to have mass. It could, for example, be a low density gas or clouds of cosmic dust. Calculations have indicated that there could well be considerably more dark matter in the universe than visible bright matter. A frequently quoted ratio of dark to visible matter is ten. If this is correct, then dark planets, like the earth, would provide too little mass to account for the required amount of dark matter. Whatever this mysterious dark matter may be, it could furnish the universal isotropic matter distribution required for Mach's principle to be valid.

Like all objects, dark matter would radiate heat. The heat waves become longer and longer as the emitting particles cool down. At a few degrees above absolute zero, the wave length ranges from millimeters to centimeters. Electromagnetic energy waves of around one centimeter length are called microwaves. In the 1960s two radio astronomers - that is astronomers who observed radio wave sources rather than light wave sources in the sky - of the Bell Telephone Laboratory were looking for cosmic background heat radiation and found it at a temperature of approximately three degrees Kelvin in the microwave band. For this discovery A.A. Penzias and R.W. Wilson [7.7] received the Nobel Prize of Physics in 1978. Even though the whole universe collaborated in producing this radiation, it was found to be very weak. Great experimental difficulties had to be overcome to isolate it from the stray radiation having its origin at the surface of the earth and even in the receiving antenna itself. On the occasion of the award of the Nobel Prize, Professor Hulthen [7.8] said:

"The task of eliminating various sources of errors and noise turned out to be very difficult and time consuming, but by and by it became clear that they (Penzias and Wilson) had found a background radiation, equally strong in all directions, independent of time of the day and the year, so that it could not come from the sun or our Galaxy. The strength of radiation corresponded to what technicians call an antenna temperature of 3°K."

Radio astronomy is often used for identifying various gas species in interstellar space. One possible interpretation of these background radiation measurements, that was not publicized at the time, was that it could represent heat radiation from a dilute dark matter distribution spread throughout all space. This hypothesis was in accord with Mach's principle and provided some astronomical evidence for distant cosmic matter which could be the origin of the forces of inertia.

Penzias and Wilson had little interest in the cause of inertia. They found a more topical and popular explanation for the existence of the microwave background radiation. In the 1960s the big-bang theory of the creation of the universe came into vogue. It was held that soon after the moment of the big explosion, matter must have been distributed quite uniformly throughout the expanding volume. Furthermore, this matter cloud must have been mixed up with a high density of free energy or electromagnetic radiation energy. Later the stars and galaxies condensed out of the matter cloud and left uniformly distributed radiation energy behind. This energy expanded into a bigger and bigger volume and, in the process, "cooled down". While most of it must still be flying into an ever enlarging cosmic expanse, some of it returns to us on earth at a temperature of 3°K. Big-bang theorists maintain, therefore, it is the wandering energy of the initial big bang explosion which finds its way into the antennas of radio telescopes. Cooled to the low temperature, it is said, the microwave background is nothing else but a relic of the explosion which created the universe.

So at the beginning of the twenty-first century we have two explanations of the remarkable discovery of the isotropic radiation background. One of them involves an isotropic dark matter distribution compatible with the *fixed star* system inherent in Mach's principle and the origin of inertia forces. The final verdict must be left to future discoveries.

Some years ago, one of us (PG) [7.9] suggested that any distribution of matter can be divided into an isotropic component on which an anisotropic component is superimposed. Inertia would be caused by only the isotropic component. This argument fails if the isotropic component contains too small an amount of matter. By looking at maps of the sky showing stars, galaxies, and clusters of galaxies, it is quite difficult to discern an isotropic matter distribution. However we are probably only seeing a small part of the universe around us.

Einstein lost some of his belief in Mach's principle when he theorized that an infinite homogeneous universe would cause an infinite force of inertia and nothing would ever be able to accelerate. This is a very valid argument, and led Einstein to conclude that the universe was homogeneous and infinite, but must exist in a finite expanding curved space. However even with his well recognized vivid imagination, he never considered the possibility that matter in the universe could be arranged in an isotropic manner in such a way that the density of the distribution decreases as you get further and further away from any vantage point. Without requiring that the earth is in a special point in the universe, this at first seems to be an impossible matter distribution. Einstein was apparently unaware of the blossoming subject of fractal mathematics which was in its infancy at his time. However Benoit Mandelbrot [7.10] and others have since demonstrated how such a distribution is quite feasible and is referred to as a fractal matter distribution. This opens up the possibility that an infinite universe, in an infinite space, acting in the manner prescribed by Mach's principle could lead to finite isotropic inertia forces. The remarkable fact that is only just being revealed by astronomical mapping of the local universe is that this fractal distribution is precisely what exists. This result is very new and therefore naturally controversial as it is not anticipated by the theories of General Relativity and the Big Bang. A more complete description of this matter distribution will be presented in the final chapter of this book.

What led Mach [7.5] to conclude that the force of inertia was dependent upon distant matter in the universe was his critical examination of Newton's views of time, space and motion. In a Scholium of the *Principia* [7.11] Newton described time as :

"Absolute true and mathematical time, of itself, and by its own nature, flows uniformly on, without regard to anything external."

In commenting on this river of time, Mach suggested that Newton was still under the influence of medieval philosophy and had neglected his own rule of basing theory entirely on observed facts. The only way we can measure and experience time is by observing relative motion between material objects. A pendulum standing on a table measures time by its swings relative to the table. Take away the pendulum and the table and there is nothing left to be measured or observed. Time has disappeared with the material objects. Mach justifiably held time to be "an abstraction at which we arrive by means of *changes of things*", concluding that an absolute flow of time would restrict us to a prescribed measurement of uniform motion. He considered this implausible when he wrote [7.5]:

"A motion is termed uniform in which equal increments of space described correspond to equal increments of space described by some [other] motion with which we form a comparison, as the rotation of the earth. A motion may, with respect to another motion, be uniform. But the question whether a motion is *in itself* uniform, is senseless."

It is difficult, if not impossible, to dismiss Mach's argument against the reality of the uniformly flowing river of time. We need worldly substances to say anything meaningful about time. All that we can ever measure are intervals between material events in units of shorter intervals between other material events which we can only presume are uniform. For instance the hand of a stop watch turns one revolution for every 60 ticks of a mechanism inside the watch. Time by itself has no practical or scientific value. In Mach's words: "It is an idle metaphysical conception."

When considering Newton's absolute motion of a body relative to absolute space, Mach commented that all our knowledge of mechanics is experimental knowledge concerning relative positions and relative motions of material objects. We have no means of detecting absolute space and labeling positions in this space. According to Mach, scientists should not be allowed to extend the empirical principles of mechanics beyond the boundaries of experience.

He accused Newton of precisely this transgression when the latter spoke of two material globes connected by a string and rotating about each other in free space. It was Newton's contention that the tension in the string proved absolute rotation, or rotation relative to absolute space. Mach pointed out that this kind of experiment, inevitably, had to be performed against the background of the fixed stars. It represented motion relative to the fixed stars. How could the great Newton have overlooked this fact? As in all relative motions, one could consider the globes to be at rest and the fixed stars rotating about them. This changes nothing in the physical setup. With either the globes or the fixed stars at rest, there must still be tension in the string caused by the inertial forces of relative acceleration, for that is what is being observed.

Mach in fact reasoned, justifiably, that it is difficult to say much about the relative motion of two bodies unless there also exists a third body to which the motion of the other two can be referred. In the words of Mach [7.5]:

"When we say that a body K alters its direction and velocity solely through the influence of another body K´, we have asserted a conception that it is impossible to come at unless other bodies A, B, C, ... are present with reference to which the motion of the body K has been estimated."

Mach went on to discuss the acceleration of two bodies toward each other as a result of their gravitational attraction. From experience, and in accordance with Newton's third law, it is known that the bodies accelerate in inverse proportions to their masses. The heavy body accelerates less than the light body. What kind of acceleration are we talking about in this case? It is certainly not the relative acceleration of the attracting bodies. which is naturally the same for both. However by Newton's 2nd law, their acceleration should differ from each other in the inverse ratio of their masses. Again a third body is needed as a steady reference in order to describe the observed motions and make them agree with Newton's laws. Common experience tells us that this third body must be a system of bodies which we call the fixed stars.

Mach [7.5] gave this brilliant summary of the interdependence of all matter in the universe.

"When we reflect that the time factor that enters into the accelerations is nothing more than a quantity that is the measure of the distances (or angles of rotation) of the bodies of the universe, we see that even in the simplest case, in which we apparently deal with the mutual action of only two masses, the neglecting of the rest of the world is *impossible*. Nature does not begin with elements, as we are obliged to begin with them. It is certainly fortunate for us, that we can, from time to time, turn aside our eyes from the overpowering unity of the All, and allow them to rest on individual details."

It is not only the force of inertia that depends on Mach's inertial system The second parameter related to the fixed stars is what Newton called the quantity of motion of a body. In modern terminology this is now called the momentum of the material object, p, defined as the mass, m, of a body multiplied by its velocity, v, that is $p=mv$. As always, the concept of velocity makes no sense unless it is expressed relative to another object. You can probably now guess what the reference body must be when we speak of momentum !

As pointed out earlier in this chapter, when a body is forced to change its velocity, and therefore its momentum, another body must undergo, simultaneously, an equal and opposite change of momentum, so that the total momentum of the universe before and after the event remains the same. For example, take two billiard balls, one red and the other white, and both of the same mass, m. If they roll toward each other with the same velocity, then after colliding they will separate from each other with the same velocity. If this common velocity before and after the collision is v, then the combined momentum of the two balls before their encounter was $mv-mv=0$. This implies that the red ball moved with the positive velocity $+v$ and the white ball with the negative velocity $-v$ in the opposite direction. The sum of the two momenta before the collision was zero. After the collision, the velocity of each ball is reversed and the sum of their momenta is again zero. Hence no change in the total momentum has occurred and as expected momentum was conserved.

In this example it was understood that the velocities were measured relative to the billiard table which stood firmly on the surface of the earth. Therefore in effect the velocities were referred to the surface of the earth. Momentum was conserved relative to the earth. In this case we may also assume that the acceleration of the earth relative to the fixed stars can be ignored. So we may argue that momentum was also conserved, at least approximately, relative to the stars of our galaxy,

We could however have chosen a different reference body as long as this did not accelerate relative to the earth. Assume a black ball, taken as a reference object, was rolling alongside the white ball and kept pace with it. Now the momentum of the white ball relative to the reference black ball is zero, because their relative velocity is zero. On the other hand, the velocity of the red ball to the black ball is twice as large as that relative to the earth. Hence we could say the red ball momentum is $2mv$. Even more importantly, the total momentum of the red / white ball system with respect to the black ball is also $2mv$.

After the collision the red ball will roll along with the black ball and its momentum with respect to the latter is zero. However the white ball, which formerly had zero relative velocity with respect to the black ball, now runs backward with velocity v relative to the table and $2v$ relative to the reference ball. Hence the combined momentum of the two balls which suffered the collision was $2mv$ before and after the collision. Momentum was therefore conserved even though the earth was not taken as the reference body.

It seems that the principle of momentum conservation holds as long as the reference body is not accelerating relative to the earth or relative to the fixed stars. But this is only approximately true and the following example will show that there can only be one unique inertial reference systems which guarantees momentum conservation.

Consider the case where a billiard ball strikes the cushion of the table and bounces back along the same path on which it approached. For a perfectly elastic collision (one in which no energy is lost as heat), the momentum of the ball relative to the table would be $+mv$ during the approach and $-mv$ after the collision. The total momentum before and after the collision are clearly not the same. Does it represent a failure of momentum conservation? Of course, what has been forgotten is that the earth, via the table, also feels an impulse and will undergo a change in its momentum relative to the fixed stars. If we take the earth as the reference system, the velocity of the earth relative to the earth is zero by definition, and no change in earth momentum becomes evident. This apparently violates the law of momentum conservation. In order to uphold this law we are compelled to use the fixed stars as the only generally valid inertial system.

Newton pointed out another very important consequence of momentum conservation. To understand this we have to consider the collision of two unequal bodies, one heavy and one light. As before, the sum of the momenta of the two bodies will be the same before and after the collision. Also at any instant the masses and positions of the two bodies will prescribe a mathematically defined point in space which is their common centre of mass, or gravity, which lies somewhere on the line between them. It turns out that momentum conservation ensures that the common centre of mass remains stationary relative to the fixed stars before, during and after the collision. This particular consequence of the principle of momentum conservation has been fully confirmed by experiments.

If the common centre of mass had been found to move relative to the fixed stars, momentum could only be conserved if matter in the distant universe had acquired velocity in the opposite direction. Because of the Machian connection between local objects and the distant universe, such a remote momentum adjustment would not seem to be impossible.

The forces of inertia which control all accelerations must be involved in these collisions which conserve the momentum of bodies on earth. However, since each of the inertia forces involves all of the universe, huge amounts of remote matter would have to yield a little because of the collision of two unequal rocks on earth. This does not seem to be very likely. It rather looks as if the immensity of the matter content of the cosmos enforces momentum conservation by fixing the common centre of mass of colliding bodies on earth with respect to the unique inertial frame of the fixed stars. Only non-local Machian interactions could bring about this remarkable result. The immediacy of momentum conservation on the billiard table leaves little doubt that we must be dealing with mutual simultaneous far-actions.

It seems therefore that Newton quantified the force of inertia and Mach interpreted it with cosmic interactions. There should now be no doubt as to the reality of these forces, particularly since both Newton and Mach derived their theories from experiments and not from hypotheses. Yet in almost all textbooks on dynamics and mechanics we still read that the forces of inertia are fictitious. Granted there are scattered warnings in the literature that the fictitious nature of the forces cannot be upheld under all circumstances. After all, what else is it that prevents the moon from falling to the earth if the centrifugal force does not exist?

Much of the confusion has arisen from assigning physical consequences to mathematical coordinate transformations. The mental leap from one set of measuring scales to another set can lead to physical fallacies. Mach realized the problems with coordinate transformations and for this reason abandoned Newton's absolute space. Returning once more to Newton's two globes which were connected by a string and revolved about each other, Newton took the tension in the string to be an indication of the existence of absolute motion. If one then chooses a coordinate system which revolves with the globes - a purely abstract mathematical operation - the globes appear to be stationary. In that case the centrifugal force should disappear and with it the tension in the string. But the outcome of experiments cannot be changed by abstract mathematical operations. Modern textbooks say that the experimentally observed tension then calls for the existence of a *fictitious centrifugal force.* Instead of questioning the reality of the centrifugal force, it would be more correct to describe the mathematical disappearing trick as fictitious, but in the current climate of theoretical physics, pure mathematics is often considered the ultimate arbiter. With Mach's principle the only plausible coordinates have to be attached to the fixed stars. Then the rotation relative to the firmament becomes the cause of the string tension.

With the introduction of Einstein's relativity theories early last century, coordinate transformations became a popular student exercise. It made the phrase *fictitious forces of inertia* fashionable. Mach wrote his mechanics book [7.5] late in the nineteenth century and consequently he had no cause to speak of fictitious forces. On the contrary, he argued that inertia forces were just as real as gravitational attraction and electric forces between charged particles.

In many books the discussion of inertia forces is limited to rotational effects including the centrifugal force. However the inertia force opposing the fall of the apple is just as real as the centrifugal force. The reason for ignoring what is the most important aspect of the inertia force, that defined by Newton, cannot be blamed on the use of coordinate transformations. Newton's inertia force faded away when Einstein introduced his theory of general relativity and removed the force of gravity from physics as well. This will be discussed in the next chapter. In his *Science of Mechanics* [7.5], Mach described a beautiful experiment which measured and demonstrated the existence of the force which opposes linear acceleration. Mach attributed the experiment to the German physics professor, Johann Christian Poggendorff. Mach's illustration of Poggendorff's apparatus is reproduced in figure 7.2.

Equal weights were suspended from the two arms of a beam balance. On the left arm the load is split into two unequal parts which are connected together by a light string running over a low friction pulley. If the pulley on the left is initially prevented from rotating, then the balance will be horizontal. When the pulley is then allowed to run free, the heavier of the two masses will accelerate downward and the lighter one will accelerate upward at the same rate. At the same time the balance indicator at the top will tilt to the right as if the sum of the two masses on the left has suddenly become lighter than the mass on the right.

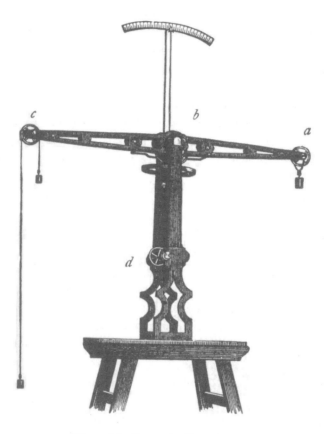

Figure 7.2 : Poggendorff's Apparatus

At first sight this may look surprising because the gravitational attraction to the earth is known to be constant and independent of the positions and motions of bodies near the surface of the earth. If only gravitational forces are acting on the balance, then the beam should remain undisturbed in the horizontal position. Yet it looks as if the two objects on the left, which, previously in a separate test, weighed as much as the object on the right, became lighter as a result of one accelerating down and the other up.

A person riding an elevator experiences an apparent increase in weight when the elevator accelerates up which he feels in his knees and leg muscles. There is a similar apparent decrease in weight when the elevator accelerates down. This should also happen to the two unequal weights on the left of figure 7.2. Since the amount of weight gain or loss is dependent on the mass of the object, the increase of weight on the lighter mass is smaller than the decrease of weight experienced by the falling greater mass. This results in a net loss of weight on the left hand side of the balance while the masses are accelerating. This experiment can only be explained by the action of real linear forces of inertia acting on accelerating bodies.

Since the two masses are connected by a taut string, the downward acceleration of the heavier mass has to be equal to the upward acceleration of the lighter mass. Let this common acceleration be denoted by a. The mass on the right arm of the balance will be denoted by $5m$, and the two masses on the left arm by $2m$ and $3m$. Then the upward force of inertia opposing the downward acceleration of the $3m$ mass is the product of this mass multiplied by the acceleration, that is $3ma$. By the same reasoning, the downward force of inertia opposing the upward acceleration of the smaller mass $2m$ is $2ma$. The two forces of inertia act in opposite directions and their net effect on the beam balance will be $3ma - 2ma = ma$. It is an upward force which counteracts the force on the left arm due to gravity. With Poggendorff's balance the inertia force can be quantitatively measured and shown to be quite distinct from gravity. This is the point that Einstein missed when making his assumptions that led to the development of General Relativity. This very real force of inertia cannot be measured on freely falling objects because the measurement procedure would interfere with the process of falling freely. That is what makes Poggendorff's apparatus so clever, for it allows the measurement of the weight of a system of masses in a modified free-fall, accelerating under the effect of gravity. It is unfortunate that there is not a diagram like figure 7.2 in every mechanics textbook because many professional physicists are still quite surprised that such a measurement is possible.

If *g* is the acceleration due to gravity of a freely falling body, then applying Newton's equations in the situation described above predicts a downward acceleration of the mass 3*m* on the balance as being equal to one-fifth of g. Hence the force which unbalances the beam comes out as *mg*/5. It is actually quite a difficult experiment to perform for practical reasons. The primary problem is that the measurable force increases with the magnitude of the acceleration. In the case just discussed, an acceleration of *g*/5 means that the heavier mass will drop 3 meters, (the height of an average room) in less than 2 seconds. This is very little time to make a good measurement with standard mechanical scales. There are two ways to get more measurement time, a) to drop the masses from a higher point and b) to make the masses less unequal. The first method certainly increases the time window. The second also increases the measurement period, but unfortunately decreases the net inertial force to be measured. For ease of description, the measurement can also be performed with the spring balance shown in figure 7.3.

If a light thread connects the larger mass (300 gm) with the hub of the pulley, as indicated on figure 7.3, no acceleration takes place. Without acceleration relative to the fixed stars, there exists no force of inertia. The balance then registers the downward force of 4.9 N (Newtons) or 500 gm dead weight. It is interesting that a term like *dead weight* should have found its way into regular usage. To follow the analogy, a *living force* would be one that is modified by the force of inertia. A freely falling apple however has no weight, *living* or *dead*.

When the auxiliary thread is cut in the spring balance experiment of figure 7.3, or burned through, as Poggendorff preferred, the weights start accelerating. This leads to a 0.196 N upward force subtracting from the 4.9 N downward gravitational force. The net result registered by the spring balance will be 4.7 N and will read 480 gm downward force, or a 4 % reduction from the *dead weight*. Disregarding small experimental errors caused by factors like vibration or air currents, any other reading indicated by the spring balance would disprove the Newton-Mach theory of inertia forces. This result however has been detected in different locations for more than a century.

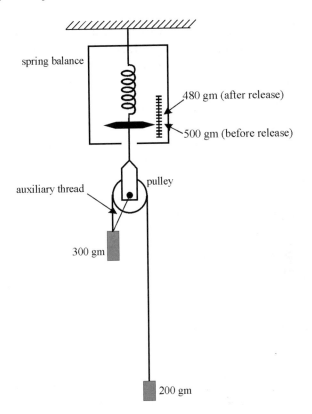

spring balance

480 gm (after release)

500 gm (before release)

pulley

auxiliary thread

300 gm

200 gm

Figure 7.3 : Spring balance for measuring inertia forces

The test just described has recently been performed in our lab in the Engineering Dept. in Oxford University. However, instead of using a spring balance which is better suited to weighing fish or a beam balance which takes time to settle, we have employed an electronic load cell. These modern devices have the capability of weighing to an accuracy of 1 part in 1000 and have a response time of a few thousandths of a second. Figure 7.4 demonstrates the successful results of these experiments. The solid line represents the theoretical percentage reduction in weight plotted against the ratio of the two masses on either end of the string. The black squares represent the measurements made in the laboratory. For instance using the masses described in figure 7.3

represents a mass ratio of 1.5, which is the situation measured by the bottom left data point in the graph in figure 7.4.

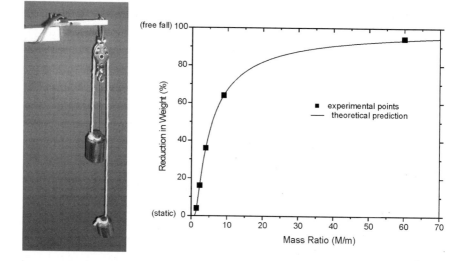

Figure 7.4 : Photo and results from recent Poggendorf inertia experiment

Strong opposition to Mach's principle has come from the large majority of physicists who believe that all action on matter must be due to direct contact between atoms or contact with the fields of remote matter. Here is a typical comment which reflects the conventional attitude to non-local inertia forces [7.12]:

"Such forces constitute the overwhelming bulk of the interactions in engineering and astronomical mechanics. Their instantaneous character flies in the face of common sense, implying that a particle here knows without delay the position and velocity of a particle there."

Mach anticipated such arguments and pointed out [7.5]:

"Like all his predecessors and successors, Newton felt the need of explaining gravitation by some means as action of contact.

Yet the great success which Newton achieved in astronomy with forces acting at a distance as the basis of deduction soon changed the situation very considerably."

Mach meant that since action-at-a-distance forces could account for gravity then they could also be the non-local connection used to explain inertia forces. He did not however foresee that the success of Maxwell's radiation theory would lead historically to Einstein's relativity and the eventual eclipse of Newtonian physics.

The inventions of radio and television, radar and lasers have been landmarks along the road of civilization, but are they more important than the accomplishments of Newtonian physics? It is difficult to see how any set of scientific rules could ever again change the world around us in such a profound manner. On this basis Mach's principle deserves equal status with field theories as possible universal schemes. Surely field action is every bit as mysterious and miraculous as distant action. Sard [7.12] argues that field contact action is common sense, implying that the human mind is incapable of comprehending action-at-a-distance. When we see a magnet float above another magnet, are you able to tell whether there is a cushion of magnetic field between them or whether they are simply repelling each other? Common sense should be admitting that we cannot distinguish easily between contact and non-local forces, but the daggers of scientific politics are too far drawn to admit such home truths.

Whatever the action principle may be, we have no answer to the question of why matter exists. Whoever or whatever was responsible for the miracle of matter creation could easily have given it properties which are no less miraculous than the existence of matter itself. What is required for mutual simultaneous far-actions is an awareness of every particle of matter of all other particles and their whereabouts. If the object of creation was to forge an interdependent universe, the awareness property of matter achieves this end with the least amount of complication.

The magic of Mach's principle is that if we lift a little finger, this action is felt by every piece of matter in the cosmos. The reach of the inertia force to the edge of the universe was not incredible at Mach's time when all the stars and planets were considered to obey Newton's universal force of gravitation. It is one thing to say a planet and the sun attract each other, and quite another thing to ascribe a force of inertia interaction between an atom and every other in the universe. Ever since the publication of the concepts behind Mach's principle in 1883, a string of accomplished scientists have tried to find the particle interaction law which must underlie it. The first of these was Albert Einstein. His reasoning and theories will be discussed in the next chapter. The effort of others is left for later. Gradually a solid Machian law of inertia is emerging. While not yet recognized by the physics community at large, its universal stature is sure to win through in the twenty-first century.

In 1987 Phipps [7.13] asked: "Should Mach's principle be taken seriously?" He truly contributed to the magic of it when he reminded us:

> "If Mach was right, the seen universe and the felt universe are two quite different places. We see because of causal (velocity of light) retardation of electromagnetic radiation (light) an old universe—a thoroughly obsolete relic. But the inertially felt universe, free of such radiation, is strictly up-to-date."

Chapter 7 References

[7.1]   B. Russell, *History of Western Philosophy.* London: Allen & Unwin, 1961.

[7.2]   R. P. Feynmann, R. B. Leighton, M. Sands, *The Feynmann Lectures on Physics.* Reading, MA: Addison-Wesley, 1964.

[7.3]   S. W. Hawking, *A Brief History of Time.* Toronto: Bantam, 1988.

[7.4]   A. Pais, *Subtle is the Lord.* Oxford: Clarendon Press, 1982.

[7.5]   E. Mach, *The science of mechanics*, 6th ed. La Salle, IL: Open Court, 1960.

[7.6]   M. Zeilik, E. P. Smith, *Introductory Astronomy & Astrophysics*, 2nd ed. Philadelphia: Saunders College Publishing, 1987.

[7.7]   A. A. Penzias, R. W. Wilson, "A measurement of excess antenna temperature at 4080 Mc/s," *Astrophysical Journal*, vol. 142, p. 419, 1965.

[7.8]   L. Hulten, in *Nobel Lectures in Physics 1971 - 1980*, S. Lundquist, Ed. Singapore: World Scientific, 1992

[7.9]   P. Graneau, "The riddle of Inertia," *Electronics and Wireless World*, p. 60, January 1990.

[7.10] Y. Baryshev, P. Teerikorpi, *Discovery of Cosmic Fractals.* New Jersey: World Scientific, 2002.

[7.11] I. Newton, *Principia (1686)*, F. Cajori (Ed.). Berkeley: University of California Press, 1962.

[7.12] R. O. Sard, *Relativistic Mechanics.* New York: Benjamin, 1970.

[7.13] T. E. Phipps, "Should Mach's principle be taken seriousiy?," *Speculations in Science & Technology*, vol. 1, p. 499, 1978.

Chapter 8

Albert Einstein

Inertia Obscured by Gravitation

The story of discovery on the path to understand inertia runs in parallel with the development of the principle of relativity. Making absolute space the cause of inertia was an unfortunate blemish on Newton's otherwise magnificent dynamics, but fortunately Mach proved quite convincingly that this stain can be removed without trouble. Newton had come close to identifying the fixed stars in the night sky as defining his absolute space. Nevertheless it remains an enigma why he preferred to hang on to the unobservable. It must be that he sensed that the stars he could observe were not distributed isotropically throughout the sky and yet inertia behaved the same way regardless of spatial orientation. While he struggled to complete the *Principia* in the 1680's, before Huygens could beat him to the punch, Newton was still thinking about the *vis insita,* his proposed inner force of inertia which was not caused by the external world. He must have found it peculiar that a particle had a gravitational force interaction with all other bodies in the universe and yet for the inertial force it was completely isolated. Newton's *vis insita* also made a mockery of his own third law for there was no matter that felt the reaction force. How could the law of gravitational attraction be so enormously successful on the basis of mutual mass interactions while the equally important inertia force shrouded itself in such mystery?

The race to complete the *Principia* and get it to the printer may have had much to do with Newton's last minute desperate decision to call upon absolute space. Both this contrivance as well as absolute time were severely criticized by George Berkeley (1685-1753) during Newton's

166

lifetime. Berkeley appears to have taken conviction from his religious beliefs and was clearly worried about the potential rise in atheism that could extend from the Newtonian scientific revolution. Berkeley's attempts to describe nature as solely composed of sensations or ideas in the mind defined an avenue of philosophy now referred to as idealism. Since it was God who produced and manipulated the images in the mind, this was considered a safe guiding principle with which science could progress without conflict with theology. He was a brutal critic of Newton's development of calculus which included the concept of infinitesimally small quantities which Berkeley considered meaningless. He could not deny the success that Newtonian dynamics had achieved in the prediction and description of moving bodies, but he was fiercely critical of ascribing occult causes to this motion such as force and gravity. Although his views on physics on the whole have not stood up to the practical scrutiny of his peers and later philosophers, he was the first to publicly point out the implausible nature of Newton's absolute space. He recognized that if the universe contained only one body, then it was simply not sensible to describe its motion. He thus deduced that we must be content with the concept of relative motion between at least two matter entities, wherever they might find themselves in the universe. One hundred and fifty years later Mach reiterated Berkeley's arguments on relativity and added them to his inertia principle.

Commemorated by the name of a university town in California, George Berkeley was educated at Trinity College, Dublin, where he became a fellow when only twenty-two years old. At that time he was a prolific philosopher of science. A large part of his important work was completed before he reached the age of thirty. Most of Berkeley's thoughts on the science of matter and motion are contained in his seminal work, *A Treatise Concerning the Principles of Human Knowledge* (1710) [8.1]. He had intended to elaborate on his concepts of dynamics in a second part to this treatise, however this was never written. Eleven years later he did produce a shorter paper entitled *De Motu* (1721) [8.1] which dealt specifically with these issues. During the latter part of his career, he travelled extensively throughout Europe and the New World. He had great hopes to set up a college on the island of Bermuda but was thwarted by down to earth bureaucratic funding problems and eventually returned to become the Anglican Bishop of Cloyne in Ireland.

Figure 8.1 : George Berkeley

Berkeley's relativity has influenced many physicists up to the present time. It is also called Galilean relativity, which is satisfied by Galilean coordinate transformations which define how events appear when viewed from moving platforms with steady velocities. These transformations simply change how events appear to others and do not affect the underlying physics. Berkeley did not obscure this fundamental principle of relativity with any further mathematical formulae. In contrast, the constant reference to coordinate transformations in modern physics has become a legacy of Einstein's theories of relativity. In the minds of many students it creates the impression that changing from one abstract viewing frame to another in space can actually change physical phenomena.

Strangely, while having been ignored by physicists for almost 200 years, Berkeley's views on the immaterial nature of reality are now firmly built into the modern relativistic theories developed by Einstein and his successors which treat energy in a field as the primary quantity of nature. Paradoxically, this energy does not consist of matter yet experimenters are required to use material devices in order to detect this energy. Modern physics has turned the tables on Newton who held that the existence of mass defined reality. In Einstein's universe, the concept of immaterial energy is now the only true reality and detectors such as telescopes and force balances are an illusion. The Bishop of Cloyne would certainly be impressed with this turn of events in the history of science.

According to his biographer Westfall [8.2], Newton abstained from commenting on Berkeley's relativity. In fact Newton may not have even become aware of it. Berkeley was not a physicist and he did not publish his ideas in the Proceedings of the Royal Society, of which Newton was president when Berkeley was writing about the irrelevance of absolute space and time. Berkeley's main treatise was published in 1710 when Sir Isaac Newton was sixty-eight years old. In the following ten years Newton was mainly involved in the administration of both the Royal Mint and the Royal Society in London. During that period most of his scientific effort was devoted to a bitter controversy with Leibnitz over who had priority in the invention of the revolutionizing mathematical tool of differential and integral calculus.

During the same period, Newton was also occupied with deep theological studies. It seems to have been religion which hardened his belief in absolute space and time as he progressed in age. He came to think that these abstract notions were related to God. In 1713, when the second edition of the *Principia* appeared, a 'General Scholium' had been added at the end of the work. In it Newton's theological convictions emerged stronger than ever. Of God he wrote [8.2]:

"He is not eternity and infinity, he is not duration or space, but he endures and is present. He endures forever, and is everywhere present; and, by existing always and everywhere, he constitutes duration and space."

So Newton's final judgment was that space and time must be parts of God. This statement has the same flavour as Einstein's later hypothesis that space and time are merged into the single concept of curved space-time.

Notwithstanding the polemics, Newton stood with his heart strapped firmly to his sleeve when he concluded the *Principia* [8.3] with:

> "And to us it is enough that gravity does really exist, and act according to the laws which we have explained, and abundantly serves to account for all the motions of the celestial bodies, and of our sea."

Since his own laws prescribed that the motions of the celestial bodies depend as much on the force of inertia as they do on gravitational attraction, he should really have stated: "that gravity and inertia do really exist". This was just the sort of atheistic materialism that Berkeley so disapproved of.

Figure 8.2 : Albert Einstein

Albert Einstein (1879-1955) really needs no introduction. He is probably the most famous physicist of all time. His career has become legendary and his relativity theories are widely known but much less widely understood. He was born in Germany, but his family moved to Italy when he was young. He eventually settled in Switzerland. Einstein received his education in Zurich and took his first job at the Patent Office. While he worked there he wrote many of his famous early papers including the development of the theory of the photoelectric effect for which he later won a Nobel prize.

Michelangelo Besso, Einstein's lifelong friend, drew the attention of the creator of modern relativity to Mach's *Science of mechanics* [8.4], while both were students at the ETH (Swiss Federal Institute of Technology) in Zurich just before the turn of the 19th century. There is little doubt that Mach's book aroused a strong interest in relativity problems in the fertile mind of the teenage Einstein. Clark [8.5] informs us that amongst Einstein's heroes, Mach ran a close second to Maxwell, the eminent Scottish physicist who founded the theory of electromagnetic fields.

These two towering scientists of the nineteenth century found themselves on opposite ends of the physics spectrum. Mach adhered rigorously to Newton's rule of *hypothesis non fingo* and the importance of reducing theory to only represent experimental knowledge. In stark contrast, Maxwell deliberately let his imagination roam. His goal was specifically to create a mathematical basis for Faraday's abstract concepts of magnetic tubes and in consequence he developed a theory of self-propagating waves flying through an undetectable ether. At the time of the publication of his theories in his famous *Treatise on Electricity and Magnetism* [8.6], there were no outstanding experiments that did not have satisfactory explanations using the existing Newtonian electrodynamics developed by Ampère, Neumann, Weber and others in Germany and France [8.7]. Maxwell consciously gave priority to mathematics even when it contradicted well established experimental facts such as Newton's third law. Strangely, rather than being shunned for his neglect of Newtonian hard nosed empiricism, he was celebrated in Great Britain under the patronage of Queen Victoria, as the modernizer who could re-establish the pre-eminence of British science. For reasons based more on politics than on experimental discovery, the Maxwellian field theory revolution eventually spread across the English

Channel and was accepted in continental Europe. As conceptual problems were eventually perceived in the man made fabric of the Maxwellian theory, an unknown Einstein discovered a mathematical trick that could save the electromagnetic field theory. In this swish of pen on paper, the ether was abolished, special relativity was born and field theory was dusted off to fight another day. In the midst of this mathematical revolution, the physics of coordinate transformations came into being. In the end Einstein became a consummate Maxwellian and his former inspiration, Mach, was pushed into oblivion.

D'Alembert's principle, which was discussed in Chapter 6, states that if a body is being accelerated with respect to an inertia system such as the fixed stars or absolute space, a force of inertia acts on it in the opposite direction to the acceleration. The force of inertia is exactly equal and opposite to the accelerating force. D'Alembert gave no explanation about how this might come about, but simply demonstrated that it always occurred. If we consider a falling apple, the gravitational force acts downward and by Newton's second law can be expressed as the gravitational mass, m_g, of the body multiplied by the acceleration due to gravity, g. The opposing and controlling inertia force acts upward and, using Newton's definition, it is equal to the inertial mass, m_i, of the body multiplied by its downward acceleration, a. D'Alembert's principle states that these two must be equal in magnitude ($m_g g = m_i a$). Since we are watching a single object, the two accelerations have to be the same, ($g = a$), and therefore the gravitational mass of the body must be equal to its inertial mass. It is in fact the experimentally proven validity of Newton's second law and d'Alembert's principle which requires the identity of gravitational and inertial mass. Newton did not take this unification for granted. Using pendulums with differing materials of equal weight, he compared hot and cold bodies, magnetic objects, and different materials such as wood, silver and gold and never found a difference between the inertial and gravitational mass to an accuracy of 1 part in 1000. In the *Principia*, he declared.

> "And by experiments made with the greatest accuracy, I have always found the quantity of matter in bodies to be proportional to their weight"

As a consequence Newton was happy in the end to simply describe objects by their quantity of matter or what we now call the inertial mass. Similar experiments performed over the successive centuries have now shown the two to be identical to at least 1 part in 10^{11}. More than two hundred years after the publication of the *Principia*, Einstein wrote [8.8] :

"We then have the following law: The *gravitational* mass of the body is equal to its *inertial* mass. It is true that this important law had hitherto been recorded in mechanics, but it has not been *interpreted*. A satisfactory interpretation can be obtained only if we recognize the following fact: *The same* quality of a body manifests itself according to circumstances as "inertia" or as "weight". In the following section we shall show to what extent this is actually the case, and how this question is connected with the general postulate of relativity."

What seems to have been obvious to Newton, d'Alembert, and Mach, but not fully accepted by Einstein, was that there exists only one kind of mass, which is a single property of matter responsible for both the forces of gravity and inertia.

The mathematics that led Einstein to his special theory of relativity (the pre-cursor to his theory of gravity) was based on several hypothetical assumptions regarding the speed of light and the time-delay of interactions between separated objects. These conjectures were based on no experimental evidence and yet to this day are considered the basis of all of modern physics. As a consequence, Einstein became the principal theorist in the movement that was criticizing Newton's action at a distance model. He simply could not accept Mach's implication of instantaneous inertia force interactions for he felt that [8.8]

"According to the theory of relativity, action at a distance with the velocity of light always takes the place of action at a distance with infinite velocity of transmission."

While Einstein was strongly influenced by Mach's opposition to absolute space, he remained skeptical of the magic of Mach's principle which stated that the whole universe collaborated simultaneously in creating forces of inertia. He did not grant that the affiliation was a force interaction. Instead he claimed it was a peculiarity of relative motion and particularly of relative acceleration between nearby bodies that we can observe and those out in the distant universe. His instincts therefore sought to find a theory that eliminated inertial forces altogether. In the course of time this Einsteinian view has led physicists to label the often powerful forces of inertia, that keep popping up in guises such as the centrifugal force, as fictitious.

Let us examine how Einstein proved the equality of the gravitational and the inertial mass of a body with his general postulate of relativity. Accepting the Berkeley-Mach proof that absolute motion is undetectable, it became obvious to Einstein that whenever one speaks of the motion of a body it must be expressed as the motion relative to one or more other bodies. This is the practical significance of the original relativity postulate. But Einstein went one step further and required that the laws of nature must be compatible with all types of relative motion between two bodies. In the original paper on the general theory of relativity [8.9] he expressed the relativity postulate as follows:

"The laws of physics must be of such a nature that they apply to systems of reference in any kind of motion."

In his most popular book on relativity [8.8], he illustrated this point by an example. In effect he considered the pressure on the feet of a person standing in an elevator which accelerates relative to the surface of the earth. We know from everyday experience that this pressure on the legs increases as the elevator accelerates upward, and decreases when it accelerates downward. If we observe the weight changes more carefully, we find there is also a weight loss while the upward motion is being reduced and a weight gain while the downward motion is brought toward a stop. According to pre-Einstein physics, all of the apparent changes in body weight are of course due to the forces of inertia aiding or opposing the normal gravitational attraction to the center of the earth.

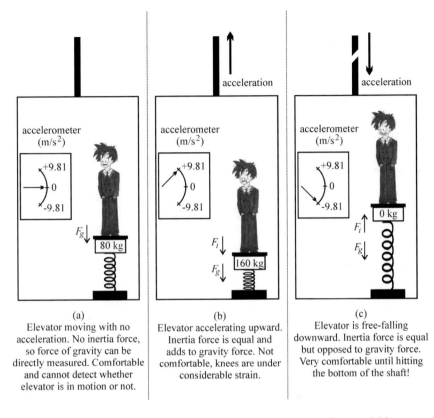

(a)
Elevator moving with no acceleration. No inertia force, so force of gravity can be directly measured. Comfortable and cannot detect whether elevator is in motion or not.

(b)
Elevator accelerating upward. Inertia force is equal and adds to gravity force. Not comfortable, knees are under considerable strain.

(c)
Elevator is free-falling downward. Inertia force is equal but opposed to gravity force. Very comfortable until hitting the bottom of the shaft!

Figure 8.3 : Demonstration of measurements in an elevator using a weighing scale and an accelerometer

The elevator experiments are depicted in figure 8.3. Inside the box we put two measuring devices, a bathroom weighing scale on which the person in the elevator stands and an accelerometer fixed to the wall. An accelerometer measures positive upward and negative downward acceleration in meters per second per second (m/s^2) relative to the Machian fixed star system. For the situation described here, we can also use the earth as an approximate reference frame. In case (a), by looking through the open door, the person standing on the scale would notice that the accelerometer reads zero when the elevator is not accelerating. At the same time the scale would register the conventional weight of the person of, say, 80 kg. If the elevator were to accelerate upward at

9.81 m/s^2 with the help of a strong motor as shown in case (b), then the weight doubles to 160 kg since the forces of gravity and inertia now add together. Case (c) demonstrates what happens if the cable is cut and the elevator then accelerates downward at 9.81 m/s^2. Due to force cancellation, there is now no net force on the scales and the person feels "weightless". From these tests the person can conclude that the acceleration due to gravity must be downwards with a value $g = 9.81$ m/s^2. This is of course the acceleration with which all bodies fall to the surface of the earth.

According to Newtonian physics, the force of inertia, F_i, opposes the force of gravity, F_g, just sufficiently to allow the acceleration of magnitude, g. This follows from Newton's second law. Hence the inertial mass of the body, m_i, incorporated in Newton's second law, must be equal to the gravitational mass m_g, otherwise the person would not feel weightless while falling at -9.81 m/s^2, and the body weight would not double when the elevator accelerates up at +9.81 m/s^2.

The Newtonian laws of nature which enter this problem are the gravitational attraction to the earth and the force of inertia opposing acceleration. These laws remain valid regardless of the various motions of the elevator. In this example the approximate system of reference was the earth. As has been pointed out before in this book, the earth cannot acquire momentum with respect to itself and therefore the important law of momentum conservation is violated. Several times in this book we have shown that the only way out of this dilemma is to use the fixed stars as the only completely valid inertial reference system instead of the earth.

Einstein however saw things differently. Since he could not abide physical laws that involved an interaction with the distant universe, he felt compelled to find some local inertial reference frames that could be used in the understanding of gravity. Special relativity had absorbed the inertial frames that had been used to describe Galilean relativity. These are frames such as shown in figure 8.3(a) in which the accelerometer on the wall reads zero. However for his theory of gravity, Einstein could not accept that frames (a) or (b) were inertial frames since in both cases there were measurable gravitational and or inertial forces. Since he had already assumed that inertial forces could not be real, he argued that in the frames shown in (a) and (b), both the gravitational and inertial forces were fictitious. Only in frame (c), where the two Newtonian forces precisely cancel could he safely say that there were no measurable gravitational forces. As a result, he defined this type of reference frame which we can call a free-fall frame as the only valid inertial frame in the theory of general relativity. He considered these frames to be special because they would be the only ones in which all of the physical laws that he considered to be correct would be valid. In order to utilise this concept, he had to conclude that all such free-falling frames of reference were equivalent. This however is simply not true.

Einstein had failed to notice that each of these frames could contain an accelerometer as shown in figure 8.3(c). This always gives a measurement of the acceleration relative to the Machian fixed stars. An elevator falling toward the surface of the moon will therefore show a different acceleration from one falling toward the earth. Worse yet, the acceleration due to gravity on earth is only approximately a constant. It is less the further one is away from the earth. Therefore, even in the vacuum of space, a box falling toward earth will be increasing its acceleration while it falls and this can be detected by the man in the elevator by watching the accelerometer dial. Since dials are controlled by forces, it is clear that the free-falling elevator has continuously changing physical conditions inside which defy Einstein's definition of a family of equivalent inertial frames. Without a set of valid local inertial reference frames, Einstein was simply not justified to formulate his relativity postulate.

Does the elevator example justify Einstein's original claim that the relativity postulate is absolutely necessary to prove the equality of inertial and gravitational mass? The answer is "no", because this equality is already embedded in d'Alembert's principle of dynamic equilibrium which always agrees with Newton's second law. Newton clearly stated that in free fall the force of inertia is equal and opposite to the force of gravitation, implying the equality of gravitational and inertial mass.

It is an empirical fact that the dynamic equilibrium between the force of inertia and any kind of externally applied force - not just gravity - will prevail whatever the motion of the body in question, as long as the acceleration of the body is expressed relative to the fixed stars. With this constraint to a unique system of reference, Mach had already discovered well before Einstein's principle of equivalence that the relativity postulate necessarily results in the equality of inertial and gravitational mass.

Einstein's theory of gravitation, which is known now as General Relativity, represents a huge intellectual achievement and its equations are far too difficult to be described here. The original paper [8.9] is a treatise on four-dimensional tensor calculus with a few glimpses of the behaviour of matter interspersed between the equations. The pre-eminent field physicists, Maxwell and Einstein, made it appear as if mathematics was the essence of physics and not simply its servant, providing the ability to make quantitative predictions. Field physics relies much more on intuitive assumptions and imaginative models than the rather drab experimental discipline of Newtonian simultaneous far-actions. General Relativity for instance describes the motions of celestial bodies as being determined by the local geometry of curved space-time rather than by attractions or repulsions along straight lines. Nevertheless, while nobody can directly detect space-time, all observers agree that freely moving bodies accelerate toward and away from each other according to Newton's laws.

The problem of creating mathematical constructs to represent reality without physical features is that it is impossible to discover whether they really exist or not. At best, one might give a probability as to the validity of any such theory. To the question of whether field physics and general relativity are true, our answer would be: probably not. At the beginning of the twenty-first century all physical theories are still in flux and continuously changing on a scale of centuries. No permanent, unflawed theories have yet surfaced and it may well take more centuries before some of the final truths about nature's ways eventually emerge.

Should researchers despair ? Certainly not ! Scientific knowledge is foremost a collection of experimental facts and astronomical observations which do have permanency. Adding new and unexpected experimental facts to this collection is highly rewarding. By all means, we should try and make sense of all the empirical facts by speculating with many kinds of theories. However it would be unwise to expect any of them to last very long, other than the Newtonian type of theory which is little more than a verbal and mathematical description of observations.

It was primarily with the general theory of relativity that Einstein created the impression that physics was primarily mathematics. Many readers have probably come to believe that mathematics is an instrument of nature. Unfortunately with all the available mathematical skills at their finger tips, physicists have no chance of making the unexpected experimental discoveries upon which successful science thrives. The doctrine of mathematical supremacy overlooks the fact that mathematics is really an invention of the human mind. Without mankind, mathematics does not exist. With fair certainty, the matter that makes up the universe was here billions of years before the human race evolved. So it is unlikely that the laws of nature were laid down so as to comply with our mathematical rules.

Electric and magnetic forces do not appear to arise from the mass property of matter, but both the forces of gravitation and inertia clearly do. This was recognized by Newton and he provided the quantitative definitions of these two mass related forces. Einstein went beyond this by combining the two mass-dependent forces into a single new phenomenon which can be called relativistic gravitation. It is certainly not the same thing as Newton's universal gravitational attraction. The elevator experiments discussed earlier were put forward by Einstein to support his unification of ordinary Newtonian forces of gravitation and inertia to create the single new concept of relativistic gravity.

Einstein argued that with the elevator door closed, the person inside cannot know if the pressure on his feet is due to the gravitational attraction of a large mass underneath him, like the earth, or whether he is feeling an inertial force caused by the acceleration of the elevator relative to a large amount of mass in the distant universe. This acceleration could, for instance, be the result of firing a rocket engine attached to the elevator. Knowing Newtonian physics and Mach's principle, the person inside might guess that the pressure on his or her feet is due to either mass-related gravity or inertia forces or a combination of both. Einstein then advises not to inquire any deeper and accept his generalised postulate of relativity since he considers that all such situations are entirely equivalent. He went further to remark that the person inside is so ignorant of the physical effects acting on the box that he can consider himself to be at rest.

However by opening the door of the elevator, the curious occupant will soon see whether he is stationary with respect to a large nearby object or is in fact accelerating. By ignoring this possibility, Einstein seems to have implied that opening the door and looking out is cheating and to some extent it is, in that it involves visual measurements which are not physical forces.

So is there a method of detecting whether one is being attracted by a gravitational force or being accelerated while trapped in a closed box without windows? In fact, there is such a method as shown in figure 8.4. In this diagram we see the occupant of the closed box subject to (a) gravitational force caused by a massive object and (b) inertial force due to acceleration. If in either case, he takes two objects and hangs them from two strings as shown, he will be able to measure the distance between the strings at the top and bottom. If the distance is the same at both places as in (b), then the strings are parallel and he must be purely accelerating far away from any significant large objects. However, if the strings are nearer to each other near his feet than at head height, then all of the objects in the box must be subject to gravitational forces which are attracting them toward a single centre of mass. To the observer in case (a), there seems to be an attraction between the two masses which is often referred to as a tidal effect. Therefore if the strings are not parallel, then he is able to conclude that the pressure he feels on his feet must be gravitational. In other words a person can indeed distinguish (with difficulty) whether he is being subjected to gravitational or inertial forces.

On the surface, this result seems to violate Einstein's principle of equivalence, however he concluded that the only valid laboratory in which to test his equivalence principle must be one that is infinitely small because then one cannot compare the motion of two separate objects. Proponents of Einstein's theory of gravitation are thus forced to conclude that the tidal effects that are observed in real laboratories are not due to a force at all, but are the result of the curvature of space. The important lesson to be learned is that for any experiment with physical size, Einstein's equivalence principle is not valid and the effect of gravity and inertia can be distinguished by careful experimentation.

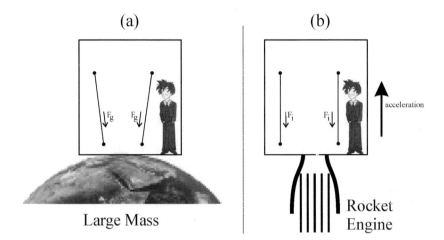

Figure 8.4 : Demonstration of "tidal effects" which violate Einstein's equivalence principle in real laboratories

According to general relativity, there is no force acting on a freely falling apple and it is the structure of space-time which determines its path. The geometry of this abstract space-time is deemed to be moulded by the sum total of mass in the cosmos. By contrast, in the Newtonian world, the direction of motion of a body is determined by gravitational attraction. In the fall of the apple, space-time is straight and guides the apple along a vertical line down to the earth. However when it comes to a planetary orbit, the path of the planet is an ellipse. Such is the curvature of space-time near the earth. So the machinery of general relativity consists of all the masses of the universe and with them it moulds the configuration of space-time everywhere in the cosmos. The curvature of space-time seems to have grooves in four dimensional space, known as geodesic lines, along which force-free matter must accelerate. The theory also contains the peculiar non-linear mechanism that a piece of matter affects the local space-time curvature around it which then in turn guides its motion. Einstein made very sure that even though it bears no conceptual similarity, general relativity agrees quite closely with the motions of bodies predicted by Newtonian mechanics.

Engineers and astronauts, who could not practice their trade without a valid and useable matter dynamics theory, rely exclusively on Newton's laws. They find this convenient and so far no spaceship has even slightly deviated from its Newtonian trajectory. Mathematicians, theoretical physicists and philosophers on the other hand, are apt to argue that general relativity is more elegant, and this is why it should be preferred to the Newtonian dynamics.

The experiments that purport to support the theory of general of relativity and demonstrate that it is more valid in some situations than Newtonian gravitation are still a matter of great debate. None of them can be performed in a controlled laboratory environment. One of the features of good scientific method taught to all students at school is that in order to test a hypothesis, one must perform two versions of any experiment. If for instance, one is testing the effect of a known mass, M, on the motion of another particle, m, one version, often called the control experiment, must remove the mass M in order to confirm that the effect on m disappears. If this is not the case then one has to go back to the drawing board and redesign a better experiment that allows a clearer identification between cause and effect. Newtonian gravity has been successfully subjected to this style of experimentation for the last 300 years. Unfortunately to test the very minute differences between Newtonian theory and general relativity, we have had to rely primarily on astronomical observations which we are unable to subject to the crucial criterion of a control experiment.

There are however several laboratory-based experiments that purport to support general relativity rather than Newtonian gravitation. These all involve the interaction of electromagnetic radiation with a gravitational potential. Newtonian gravitation, used in the manner presented in this book, only deals with the forces due to mass alone and takes no account of electromagnetic mass which is a feature of Einstein's theory of special relativity. Consequently experiments that involve no moving masses such as the measurement of the gravitational redshift are no test of Newtonian gravitational theory. As a consequence, both theories can still be said to agree with all known scientifically controlled experiments.

Does general relativity agree with Mach's principle? In the beginning Einstein sought to ensure that the answer was yes. To explain inertia Mach had brought the distant masses of the universe into the picture and Einstein claimed that his relativistic gravitational field has all the masses of the universe as its source. This should have satisfied Mach's principle. However Einstein's field equations were also fundamentally based on his proposed universal equivalence of gravitational and inertial forces. This suggested to him that the two forces can always be combined to create zero net force, and as a consequence he deemed that there were no such things as gravitational or inertial forces. The mathematics that allow this amalgamation also produce the geodesics (lines of space-time curvature) which control the direction and acceleration of moving bodies and distinguish general relativity from Newton's universal gravitation.

Much to his dismay, Einstein found in the end that his removal of the forces of gravity and inertia did not comply with Mach's principle. In 1917, an eminent Dutch astronomer, Willem de Sitter, pointed out to Einstein that there was a finite valued solution of his field equations that gave the inertial mass of a particle even if it was the only one in the universe. In this case, the curved space-time of general relativity would be flat, that is the geodesic line passing through the particle would be straight. The lone particle would be guided along this geodesic line as if it was made of inertial matter. Einstein initially argued strongly against this solution. However eventually he conceded that his interpretation of inertia could not therefore be due to other matter, as required by Mach's principle, because there was no other matter around in the de Sitter example. Thought experiments of this kind were Einstein's trademark and ultimately his theory forced him to divorce himself from the brand of relativity of his former source of inspiration, Ernst Mach.

Einstein's principle of equivalence states that inertia forces cannot be distinguished from forces of gravity. This is, however, not true for all circumstances in which inertia forces arise. When a magnet attracts another magnet, the magnetic force of attraction or repulsion will also be opposed by an equal force of inertia which controls their relative acceleration. The same applies for electric forces. Neither magnetic nor electric forces are based on mass and thus neither could be equivalent to the resulting inertia forces. The involvement of inertia forces in the dynamics of electromagnetism is the primary reason why Einstein failed to find what he called a "unified field theory".

Some kind of unification of electric and magnetic forces was present in Maxwell's electromagnetic field theory. If this could have been extended to the inertia-gravity field of general relativity, Einstein's ambition would have been fulfilled. As his biographer Clark [8.5] points out, Einstein's lifelong dream of creating the unified field theory began soon after the publication of general relativity in 1915. It took fourteen long years until the first paper appeared just before his fiftieth birthday which he still celebrated in Berlin. His theoretical problems were characterized by Wolfgang Pauli, a much younger man who made his name in quantum theory, who said of Einstein's unification effort [8.5]:

"What God has put asunder, no man shall ever join".

In an interview published in the Daily Chronicle a few days after the first unified field theory saw the light of day in January 1929, Einstein is reported to have said:

"Now, but only now, we know that the force which moves electrons in their ellipses about the nuclei of atoms is the same force which moves our earth in its annual course about the sun, and is the same force which brings to us the rays of light and heat which make life possible upon this planet."

Within a year Einstein abandoned the first unified field theory and made a fresh start with the Austrian physicist Walther Meyer. In October 1931 the details of a new unified field theory were made known. It did not fare any better than the first. Einstein said that these and subsequent attempts had caused him: ".... an agony of mathematical torment from which I am unable to escape." As mentioned in Chapter 1, the depth of depression caused by his failure of unifying electromagnetic and gravitational field theories, made Einstein write to his old friend Maurice Solovine in 1949 that his life's work may not "stand firm".

Nevertheless it is still fashionable to work on force unification although it is not at all clear what can be gained from this theoretical exercise. Are we to believe that the Creator had some incentive to use only one kind of force, or law of nature, to make the miraculous universe tick? Was it His purpose to make it easy for man to understand Nature? Unification brings unquestionable advantages to the teaching of physics. What could sound more impressive to a young student than to be told that all knowledge flows from a single abstract concept. Most of all, force unification would add to the aesthetics of science. None of these noble aims make unification necessary or inevitable.

There is a more negative aspect to the search for unification. The discovery of the rules of action prevailing in the physical world have ultimately enabled us to create the civilization in which we live. Man's separation from the animals has come about through the power of invention which has culminated in the technologies of recent generations. The inventor has to be able to predict and visualize what will happen when certain things are put together in a certain way and supplied with energy. To do this he calls upon much scientific information, some experimental and some theoretical. The more varied his information and theories are, the better are his chances to find solutions to new challenges. If unification means the combination of different scientific fields into one subject, it will inevitably shrink the armoury of the inventor.

Strangely, there is some connection between Einstein's theory of gravitation and Maxwell's electromagnetism. Electromagnetic waves - that is light - are supposed to be deflected in Einstein's gravitational field. This connection arose from Einstein's theory of special relativity in which a Maxwellian electromagnetic wave is said to possess mass, not matter-mass but equivalent electromagnetic mass. Hence light should be attracted to heavy objects or, to be more precise, its path should be curved near massive bodies. The predicted deflection was so small that it could not be observed unless the light grazed the surface of something as large as the sun. To check this prediction, one could look at stars further away than the sun, but visible at the sun's periphery.

Because of the brightness of the sun, this observation could only be made during a total eclipse of the sun. The first opportunity arose on May 19, 1919, when a British expedition to Principe Island performed the telescopic measurements. The leader of the expedition was the British astronomer and relativist Arthur Eddington. It took almost six months before the members of the expedition could agree that they had found the small effect predicted by general relativity. They announced their findings on November 6, 1919, to the assembled Fellows of the Royal Society and the Royal Astronomical Society of London. The next day the London Times hailed the success of the expedition with an article headed "The fabric of the universe". Einstein woke up that morning in Berlin to find himself famous. No other scientist had ever received such instant worldwide acclaim.

While Newton piled up example upon example of the application of his laws to celestial and earthbound mechanics, Einstein's general relativity paper professed little added utility. As Newton's theory of gravitation was confirmed by reams of astronomical observations and laboratory data, general relativity had to agree with it closely.

We know of only one example to which both theories apply and on which they disagree. This concerns the slow rotation of the perihelion of the elliptic orbit of the planet mercury. The perihelion of a planetary orbit is the line of the shortest distance from the planet to the sun. Careful astronomical measurements have shown that the perihelion line of every planet rotates slowly with respect to the fixed stars, The rate of perihelion rotation for mercury is of the order of one revolution per 225,000 years. For all planets, except mercury, the advance of the perihelion has been explained with Newtonian gravitational interactions between the planets. In the case of mercury the observed advance of the perihelion comes to 575 arc-seconds per century. There are 1,296,000 arc seconds in one complete revolution. All but 43 of the 575 are accounted for by the Newtonian gravitational attraction between the various planets of the sun. The small discrepancy for the mercury perihelion has lead astronomers to search for unknown matter in the solar system since the middle of the 19th century. None has been found. This left the door open for a general relativity explanation which Einstein duly provided. It is said to have been the happiest moment of his life when he discovered that his calculations agreed with the long standing astronomical anomaly. It should be pointed out that the 43 arcsecond/century has since found other non-relativistic explanations as reported by Phipps [8.10], Assis [8.11] and Rowlands [8.12]. As a test of general relativity, the astronomical measurement still suffers from not being a controlled laboratory experiment and thus cannot be used to dispute the continuing validity of any theory including Newtonian gravitation.

Unlike many of his followers, Einstein was very much aware of the continued success that Newtonian gravitation and inertia have enjoyed. In his book *Out of my later years* [8.13] he wrote:

"No one must think that Newton's great creation can be overthrown in any real sense by this (general relativity) or by any other theory. His clear and wide ideas will forever retain their significance as the foundation on which our modern conceptions of physics have to be built."

Unfortunately as a consequence of his flamboyant and highly publicized career, Einstein's followers were already committed to a scientific path on which Newton was considered yesterday's man. Not for the first time, the history of science witnessed the removal of a successful empirical theory from the ranks of acceptability simply to make way for a mathematical scheme which felt more modern and intellectually exciting. Inertia may temporarily have been obscured in the textbooks by Einstein's theory of gravitation, but the evidence discussed in this chapter has shown that it is a very real phenomenon. The failure to unify the theory of general relativity with other aspects of physics means that the force of inertia and Newtonian dynamics must make an inevitable comeback in the minds of physicists.

Chapter 8 References

[8.1]   G. Berkeley, *The Works of George Berkeley*, A. A. Luce and T. E. Jessop (Eds.), vol. 2&4. Edinburgh: Thomas Nelson, 1948-57.

[8.2]   R. S. Westfall, *The Life of Isaac Newton*. Cambridge: Cambridge University Press, 1993.

[8.3]   I. Newton, *Principia (1686)*, F. Cajori (Ed.). Berkeley: University of California Press, 1962.

[8.4]   E. Mach, *The science of mechanics*, 6th ed. La Salle, IL: Open Court, 1960.

[8.5]   R. W. Clark, *Einstein, the life and times*. New York: World Publishing, 1971.

[8.6]   J. C. Maxwell, *A Treatise on Electricity and Magnetism (1873)*. New York, N.Y.: Dover, 1954.

[8.7]   P. Graneau, N. Graneau, *Newtonian electrodynamics*. Singapore: World Scientific, 1996.

[8.8]   A. Einstein, *Relativity: The special and the general theory*. New York: Crown Publishers, 1961.

[8.9]   A. Einstein, "The foundation of the general theory of relativity," in *The principle of relativity*. New York: Dover, 1923

[8.10] T. E. Phipps, *Heretical verities: Mathematical themes in physical description*. Urbana, IL: Classic Non-Fiction Library, 1987.

[8.11] A. K. T. Assis, "On Mach's Principle," *Found.Phys.Lett*, vol. 2, p. 301-318, 1989.

[8.12] P. Rowlands, *A revolution too far*. Liverpool, UK: PD Publications, 1994.

[8.13] A. Einstein, *Out of my later years*. London: Thames & Hudson, 1950.

Chapter 9

Inducing Inertia

An Electromagnetic Analogy

Even after the unprecedented rapid international acceptance of Einstein's theory of general relativity, there was surprisingly no immediate upsurge in the number of physicists who devoted themselves to its further development. This seems to have been a consequence of the theory's complexity and also the technical difficulty of performing further experiments with which to test its subtle divergence from Newtonian gravity.

In the first few decades of the twentieth century, there seems to have been a healthy sense that theoretical physics should not stray too far from what can be measured by controlled experiment. The development of the theory of quantum mechanics can be cited as an example of theory that always followed one pace behind unexpected laboratory discovery. However, after World War II, there was a revolution in the scientific world as it became perceived that physicists had played a highly significant role in determining the war's eventual outcome. With such encouragement and significant increases in funding, whole new areas of research came into being and some old ones were rediscovered. Such a period of renewed confidence had the effect that pure mathematical research became much more acceptable than it had been before the war. The mathematical development of gravitational theory was one of the subjects that accelerated tremendously in the early 1950's.

A young PhD student at Cambridge University named Dennis Sciama (1926-1999) was one of this new breed of mathematicians who was keen to explore how far one could push the theory of general relativity. As a result of his contact with Paul Dirac, one of the most famous founders of quantum mechanics, Sciama had also developed a fascination with inertia and Mach's principle.

Figure 9.1 : Dennis Sciama

Early in his career, Sciama produced one of the most frequently quoted inertia papers, "On the origin of inertia" [9.1]. This was published in the Monthly Notices of the Royal Astronomical Society of 1953. He began his argument with Einstein's confession that general relativity did not deal satisfactorily with the inertial behavior of matter which seemed to open the door for a revisitation of Mach's principle. As well, Sciama was certainly aware of Einstein's disappointment that he had been unable to incorporate Mach's principle into his theory of gravitation. It was therefore a brave physicist who in the middle of the twentieth century dared to pursue a Machian theory of inertia. At the same time it was inconceivable to disregard Einstein's thoroughly accepted field theory simply in order to explain the inertial force.

Sciama belonged to Trinity College, Cambridge which for many years had been home to Newton. However even such powerful ghosts drifting through the hallowed cloisters could not influence Sciama to promote a return to Newton's action at a distance philosophy. So Sciama set about to attempt some repair work on field theory which was meant to leave general relativity intact. He conceived that inertia was induced in matter, rather than being an inherent property of material substances, having borrowed the induction concept from the subject of electromagnetism.

Electromagnetic induction has a precise meaning in the action at a distance electrodynamics of Coulomb, Ampère, Neumann and Kirchhoff [9.2]. In Maxwell's field theory [9.3] it leads a more loosely defined existence. After Faraday's famous discovery of electromagnetic induction in 1831, it was Franz Neumann who first subjected this phenomenon to mathematical analysis. Neumann formulated two laws of electromagnetic induction, neither of which were inverse square force laws of the type that had been previously applied to Newtonian matter interactions. Neumann's laws did not predict ponderomotive (mechanical) forces, but dealt with something which Neumann himself called 'electromotive force'. It has turned out to be an unfortunate choice of words which is still with us 160 years later. In all of physics except electromagnetic induction, the term 'force' is reserved for mechanical action on ponderable matter and is measured in units called Newtons. In contrast electromotive force is measured in units of Volts and causes charge separation.

Maxwell admitted freely that he did not understand the connection between electromotive and ponderomotive forces which were both produced by electric currents. He warned readers of his treatise [9.3] not to confuse the two types of force and said:

"Electromotive force is always to be understood to act on electricity only, not on the bodies in which electricity resides. It is never to be confounded with ordinary mechanical force, which acts on bodies only, not on the electricity in them. If we ever come to know the formal relation between electricity and ordinary matter, we shall probably also know the relation between electromotive and ordinary force."

Andre Marie Ampère, whom Maxwell described as the Newton of electricity was the first to discover a ponderomotive force law between separated sections of electric conductors in 1822. He described these sections of metallic conductor as current elements and his law described a force of attraction or repulsion between them that was expressed in units of Newtons and had the same mathematical structure as Newtonian gravitation. However instead of mass, Ampère's electrodynamic force depended on the strength and direction of the current flowing through the element. He developed this law while in complete ignorance of the microscopic structure of a conductor which we now describe as a sea of free electrons moving through a fixed lattice of stationary ions. When Faraday discovered the phenomenon of electric induction in 1831, he was well aware that he had not found a Newtonian force that caused conductors to move. Rather he had stumbled upon the conditions by which the motion of conductors or magnets or a change in current strength can affect the voltage and current in nearby conductors.

The two types of force differ not only dimensionally, but also with respect to their reaction forces as well as the distance over which they are effective. For any particle of matter which is exerting a ponderomotive force on another particle, Newton's third law requires that the other particle must exert an equal and opposite reaction force back on the first one. The induced electromotive force has no such reaction force. It is an action caused by one atom on another atom, but the second atom has no means of reacting back on the first atom. In other words, the electromotive force is neither an attraction nor a repulsion. It is a one-way effect.

Newton's Coulomb's and Ampère's ponderomotive forces between gravitating particles, electric charges and current elements respectively, not only obey Newton's third law, but their strength falls off with the inverse square of the distance. The electromotive force decreases more slowly and is only inversely proportional to the distance of separation. This is the reason why electromagnetic radiation is so effective for long-distance communication. If inertia is analogous to electromagnetic induction as Sciama proposed then it should have these two properties: (1) one-way action and (2) long reach.

In order to avoid the philosophical pitfalls of absolute motion, Sciama used Mach's principle as a guide to construct his theory. His motivation stemmed from a desire to marry the non-local instantaneous interactions implied by Mach's principle with the local field physics of Maxwell and Einstein. He hoped that by the application of mathematics, these conflicting action principles could achieve peaceful coexistence.

By the 1950's, astronomical knowledge of the universe had advanced considerably since Mach's time. It was now clear that our sun is just one of billions of stars that make up our disk shaped galaxy and more importantly that there are an inestimable number of galaxies distributed all around us. With this new information, it is interesting to observe how Sciama convinced himself that the matter responsible for Machian inertia had to reside well beyond our Milky Way home galaxy. To this end he reasoned that even though our galaxy is not homogeneous and isotropic, it was likely that the more distant matter in the universe could be considered to be distributed with constant density. He therefore decided that the amount of mass in any given spherical shell centred on the earth must increase roughly with the square of the distance from the earth. Then he hypothesized that the inertial influence that cosmic matter could exert in our laboratories decreased more slowly than the inverse square of the distance. If this was true, the most distant matter in the universe became of predominant importance and the effect of nearby objects like the sun became insignificant. Hence Sciama's long range forces of inertia could be represented by similar equations to the long range electromotive forces of induction.

For the distant matter distribution, Sciama chose a homogeneous and isotropic model. This distribution did not change with time and, therefore, to everything in our galaxy it looked like being stationary over the timescales of our observations. This remote and homogeneous matter distribution obviously represented Mach's inertial reference frame which the Austrian physicist called "the fixed stars". It was a unique inertial system, defined by real observable objects and could not be replaced meaningfully with any other inertial frame.

Hoping to merge Mach's principle with general relativity, it was Sciama's contention that the forces of inertia were forces of a gravitational nature. This was however not the same as Einstein's principle of equivalence, in which it was asserted that the gravitational and the inertial forces were equivalent and always cancelled each other out and therefore did not exist. In contrast, Sciama clearly wanted to demonstrate that gravity and inertia were related, but nevertheless separate, mechanisms. This would explain why inertia forces are isotropic (direction independent) due to the distant homogeneous universe and the gravitational forces of which we are aware are highly anisotropic due to the irregular distribution of matter in our local environment (solar system and Milky Way). It is interesting that no gravitational force law was spelled out in Sciama's paper. He could not have been using Newton's law of universal gravitation because the author claimed that his inertia-gravity force was retarded and travelled between objects at the velocity of light.

This leads to an immediate difficulty with the inertia-gravitation field model. Take Mach's principle in the popular form which Phipps [9.4] attributes to Mach himself.

"When the subway jerks, it's the fixed stars that throw you down."

Understanding this chain of events requires a little careful thought. The moment that the train driver applies the brakes, the subway traveler feels a backwards force applied through the soles of his feet. This force is then transmitted through the structure of his body so that his head and shoulders also feel the deceleration. As a consequence of this deceleration with respect to the fixed stars, there must also be a force of inertia which now acts in a forward direction on every atom in his body. While his feet and shoes will generally stay fixed to the floor of the train by friction, his upper body is more free to respond to the force of inertia causing him to fall forward with respect to his feet. We know that this is a real force because we can actively try and resist it by tensing several muscle groups to try and stay upright.

How can this immediate effect be the result of an interaction with the fixed stars, or more precisely, with uniformly distributed matter in the remote universe ? Retarded interactions will never be able to account for the immediate appearance of inertia forces because it would take a significant and possibly infinite period of time to transmit the information of the subway braking to the distant universe if limited by the speed of light. If the matter in the universe does not receive a net backward acceleration at the same moment that our subway rider's head is pushed forward of his feet, then momentum will not be instantaneously conserved in the universe. As far as we are aware, nobody has ever proposed that the universe as a whole can gain or lose momentum. Therefore the distant universe must be immediately aware that the train driver has applied the brakes and must react to all of the consequences. Mach's principle can therefore only be compatible with simultaneous distant matter interactions of the type described by Newtonian gravitation.

Notwithstanding this inevitable conflict, Sciama persevered with the retarded energy transport of field physics. It became a mathematical puzzle. Many of the most famous theoretical physicists of the late twentieth century have professed that advancing our understanding of nature was simply a matter of juggling mathematical equations until something significant was found which agreed with experiments and the currently prevailing philosophy of science. Sciama's attempt was no different. In the end, he was able to come up with equations that yielded the force of inertia when he fed in the best estimates of the day for the size and density of the universe. However his theory would always be criticized for being neither consistent with general relativity nor Newtonian physics and thus always fell between two stones. Consequently, although often referenced, it never really acquired a following.

He had set himself the Machian goal to show that matter has inertia only in the presence of other matter. He thought that this criterion required an inertia law of induction. To understand the difficulties that blocked his way, it helps to look once more at electromagnetic induction. Starting with Maxwell's equations, as was Sciama's declared analogy, electromagnetic induction is defined by what is usually called Faraday's law. This states that the electromotive force induced in a closed circuit is equal to the rate of change of magnetic flux linkage through the circuit. Unfortunately nothing in the realm of gravitation and inertia can be equated to a 'closed circuit'. It forced Sciama to abandon Faraday's law, and with it Maxwell's equations - although he never admitted it - and to define the induced electromotive force at a point in terms of the magnetic vector potential. These potentials are actually a relic of Newtonian electromagnetism [9.2]. They do not involve field energy which has equivalent electromagnetic mass and is essential for Einstein local field contact action. We should not be surprised that physicists and electrical engineers still use a handy mathematical tool even though it does not appear to fit the theory that they believe to be true. Astronomers and spacecraft engineers regularly still use Newton's law of gravity instead of the much more unwieldy general relativity.

There is however another theory which cannot be formulated without the magnetic vector potential which is quantum mechanics. This was first pointed out by Aharonov and Bohm [9.5] in 1959. They predicted that there would be a force on an electron in a region where no magnetic field exists. However, at the position where the electron is observed to be accelerated there is a magnetic vector potential. Experiments have been confirming this phenomenon since the early 1960's [9.6]. This fact, now called the Aharanov-Bohm effect, as well as many others since, have proved that quantum theory has to be based on Newtonian action at a distance. However, so as not to close the door on a hopeful reunification with field theory, quantum mechanics is described as a non-local theory and the words *action at a distance* are never uttered. For readers who think that political correctness has not entered the realms of pure physics, they are sadly mistaken. So to put it directly, Quantum mechanics, Newtonian physics and any inertia theory which is designed to be consistent with Mach's principle will inevitably have to be non-local action at a distance theories and cannot be consistent with modern electromagnetic field theory.

Sciama's gravitational vector potential was closely associated with Newton's law of universal gravitation. Quite reasonably, he felt only able to consider a homogeneous and isotropic matter distribution in the remote universe for there was no evidence of any other arrangement. This naturally led to a net gravitational force and gravitational vector potential on a point particle here on earth of zero magnitude because of symmetry. In other words, the forces pulling a particle in all directions balance out. At first, this seems to rule out inertia induction by the gravitational vector potential.

Sciama overcame this obstacle, by using Hubble's model of the expanding universe. In the 1930's the astronomer Edwin Hubble [9.7] had discovered that the frequency of light emitted by virtually all remote galaxies was shifted toward the red end of the spectrum. This was considered to be the Doppler effect of a receding light source. This is the electromagnetic equivalent of the acoustic mechanism that makes the sound of a car a lower frequency when receding from you than when approaching. As a result of Hubble's measurements, it became widely accepted that virtually all observable matter in the universe was moving away from us. It meant that the universe had to be expanding. Hubble also discovered that the amount of red shift on the spectral scale increased the further the galaxy was removed from the earth. These findings culminated in the development of the Big Bang Theory.

To make this point Sciama said

"We shall assume that matter receding with velocity greater than that of light makes no contribution to the potential (on earth), so that the integral (of the potential) is taken over a spherical volume."

This is certainly not true in Newtonian physics in which there is no velocity of light limitation to gravitational attraction. Newton himself rejected the idea of a universe of finite size for he thought it would fall inward and collapse into a large spherical lump [9.8]. On the other hand, relativistic field theory, to which Sciama wished to adhere, required that potentials emitted by a piece of matter travelled at the velocity of light. This led to the possibility that the universe may be infinite in size, but that there was only a finite amount of matter inside this sphere that could affect the physics on earth. The radius of the sphere was $c\tau$ where c is the speed of light and τ is the time since the Big Bang. Such a scheme certainly produced a solution to the so called gravitational paradox which had worried many of Sciama's predecessors in which an infinite homogeneous universe leads to infinite and therefore undefined gravitational forces. However the philosophical price that Sciama had to pay was to accept that an infinite amount of matter was travelling away from us faster than the speed of light. Unfortunately under the assumptions of special and general relativity, no such motion is possible. This issue was not addressed.

Disregarding these difficulties, Sciama reasoned that the gravitational vector potential is caused by a spherical universe of uniformly distributed matter with the surface of the sphere expanding at the velocity of light. At any point near the center of this spherical universe the vector potential vanished on account of the symmetrical matter distribution and its symmetric expansion. Then he imagined a particle in the laboratory on earth, in the center of the sphere, and let it travel at constant velocity relative to the distant cosmic matter. It would then have a lower relative velocity with respect to the distant matter in front of it compared to the matter behind it. Therefore this particle would "feel" a finite gravity vector potential in the direction of motion. If the particle underwent acceleration and changed its velocity, this vector potential would also be subject to a rate of change. In electromagnetism a time-varying magnetic vector potential is always the cause of an induced electromotive force. For this reason Sciama predicted that the time-varying gravitational vector potential, which was a function of particle acceleration, was the cause of an induced inertia-gravity force. In other words there is a local force acting between the particle and the changing field surrounding it.

The fallacy of the argument is this. Any change in matter distribution on the earth will, in field theory, not be communicated to the outer cosmos for millions of years. Hence on a time scale of seconds, during which the force of inertia topples the subway traveler, there is no reaction effect on the matter distribution in the spherical universe. Hence momentum cannot be conserved at all times.

Sciama promised that certain mathematical procedures in his treatment of the origin of inertia would be improved in a more complete second paper. As in general relativity, he was going to upgrade his vector algebra with four dimensional tensor calculus. This second paper never appeared in print.

Although the concept of inertia was already appreciated in Kepler's dream, and is now almost 400 years old, we still struggle with the words that correctly describe it. Most misleading is the phrase *objects possess inertia*. This immediately conveys the idea of inertia being a property of matter, like mass. In this vein, Sciama had been guided by his statement that matter has inertia only in the presence of other matter. It was an improvement over the pre-Machian notion that inertia was fundamentally possessed by matter. However at the same time, in Sciama's theory, induction only led to a non-Newtonian justification of the force of inertia. Mach's principle is actually quite different and much more specific. It describes matter interactions, that is forces, acting directly between particles. It is the intrinsic attributes of the particles that determine the magnitude of these forces. The physics of forces and induced phenomena are quite distinct in the Newton-Mach paradigm. It is only when speaking of the forces of inertia that the meaning of Mach's principle achieves clarity.

Induction is an important scientific term. Its electromagnetic connotation is to make something happen or to change the state of another body. One object can cause the separation of electric charges or the flow of electric currents in another. However induction does not cause the two objects to move with respect to each other. As a result, Sciama was never really justified to speculate that gravitational forces could be induced.

The most important consequence of Sciama's paper [9.1] was that, in the middle of the twentieth century, it re-opened the search for the origin of inertia. Little had been said about the subject since Mach addressed it, late in the nineteenth century, and Einstein intermingled inertia with gravity in 1915. Later Einstein had realized to his dismay that general relativity did not conform with Mach's principle. However Einstein's followers had invested too much in understanding his theory to give up on it so quickly. Sciama did well to recognize that a lack of an inertia explanation was a serious flaw in general relativity. Since Sciama's paper, there have been reminders in the literature of Mach's principle which have become more frequent. An anonymous reviewer of one of our recent papers [9.9], published in the journal of *General Relativity & Gravitation* wrote

"The work submitted, devoted to the question of Machian inertia, an argument that is still opened and very significant in fundamental physics, is interesting."

The paradox of Sciama's thinking was that, on the one hand, he found it difficult to see why the principle of gravitational and inertial equivalence should be true simply by assumption as Einstein had declared. He therefore purposefully used separate expressions for the two forces. On the other hand, he continued with general relativity which could not exist without the equivalence principle which was the starting premise of Einstein's theory and its primary content. Without it there was hardly anything left of general relativity which Sciama was attempting to repair. Can a car without an engine be repaired? We can now begin to see why most of the subsequent investigators of the origin of inertia, who were stimulated by Sciama's paper, have turned away from general relativity and sought a solution in action at a distance methodology.

One aspect of Sciama's analysis that surprised him was that he was forced to predict that there was much more matter in the universe than what was being observed. His theories predicted a universal density that was many orders of magnitude higher than current estimates from telescopic observations. It predicted 5000 times as much matter in the universe than estimated by astronomers from their telescopic observations. Although Sciama's figure is likely to be in error, there is now general agreement for the resolution of several astronomical anomalies that much more matter must reside in the universe than visible bright matter. There could therefore be a large amount of dark matter which is cold and emits low temperature radiation. In Machian inertia theories it is not sufficient simply for this matter to exist. It must also be distributed isotropically throughout the universe so that the force of inertia on a given body is the same for all directions of acceleration. The microwave radiation which reaches us from all directions, as mentioned in Chapter 7, discovered by Penzias and Wilson [9.10] could well come from the remote cold dark matter which is so essential for the origin of inertia forces. Possibly Sciama's most important contribution to inertia science was his prediction of an unexpectedly large amount of matter in deep space.

With hindsight, Sciama's philosophical dilemma is a perfect illustration of the natural inconsistency between on one hand, the Machian instantaneous direct interaction of distant particles and on the other, the field theory model of bodies interacting by the radiation and detection of travelling field momentum and energy. It is remarkable that modern physics has become completely dominated with the retarded actions which Sciama took for granted. In this picture, not only must this energy momentum field travel through space, possibly to never interact with any other particle, but it is believed that it must propagate at a very specific velocity, commonly called the speed of light. There exists no direct experimental proof that any one of these assumptions is true. The belief in retarded actions is like the belief in astrology – possible but not demonstrable.

Let us consider, in turn, the two principal assumptions underlying retarded actions: (1) the existence of free energy, and (2) the unique velocity of energy travel. Where is the evidence that something like free energy actually exists in otherwise empty space? If energy is being transported from one place to another, it should be possible to set up barriers to intercept this energy movement. So far it has proved impossible to shield objects against gravitational and inertial forces. From a mechanistic viewpoint, this should be sufficient to rule out the retarded inertia actions proposed by Sciama. General relativity also predicts the existence of gravity waves which should travel through empty space. However despite many millions of dollars of research expenditure, none have been detected. Moreover as described in Chapter 1, we also know, thanks to the calculations of LaPlace and more recently by Van Flandern, that if gravitating bodies interacted with a delay represented by the speed of light, then the solar system would quickly become unstable which is not what we observe. The high accuracy of modern astronomy has determined that the speed of these gravitational interactions must be so fast that if energy is indeed travelling, it must be moving at twenty billion times the speed of light. This clearly does not fit the restrictions presented in Einstein's theories of relativity. So in fact, the only interaction mechanism that can fit all of the known facts regarding gravity is instantaneous action at a distance.

The situation is different in electromagnetism. In this case we can set up screens which will block light and other electromagnetic radiation. However this alone does not prove that electromagnetic waves exist. In this case one can only show the impossibility of the transfer of energy and momentum by electromagnetic fields with a quantitative experiment. This reasoning which is rigorously presented in [9.2] involves the force on a railgun armature as shown in figure 1.3 in Chapter 1. The experiment is in fact nothing more than an example of the electromagnetic mechanism that exists in every electric motor. The essence of the argument is built on the measurement of the mechanical momentum delivered to a metal armature by an electrodynamic force. If the force is transmitted by a travelling field then we also know how much momentum must have existed in the electrodynamic field before it collided with the armature. Modern relativistic field theory has a very specific formula which relates how much energy is contained in a field which also carries a certain momentum. So therefore we can ascertain a minimum estimate of how much energy must be in the field to provide the measured momentum. Calculations have shown that this energy needs to be up to a thousand times as large as the energy that is available for any given experiment. In this case field theory clearly violates energy conservation on numerical grounds.

By common consent, the conservation of momentum and energy are more fundamental in physics than the dogma of field theory. Therefore the railgun armature and all electromagnetic motors cannot be driven by field energy impact. The conclusion is that travelling electromagnetic energy simply does not exist. The operation of electromagnetic devices can therefore only be explained with a Newtonian electrodynamics [9.2] based on simultaneous mutual far-actions. Machian inertia forces require no less.

In 1998 we mailed an early draft of this chapter to professor Sciama in Oxford. In his response Sciama said:

> "If you jerk your head back you immediately see a change in the
> red shift of a distant galaxy, although its radiation is propagated
> at the speed of light."

Presumably he meant that when the subway jerks, the gravity field of the distant stars can still throw the traveller down despite the fact that energy travels at the speed of light, demonstrating his continuing conviction in the tenets of general relativity. Nevertheless, the fact that his response to questions regarding gravity and inertia should still rely on the red shift of light demonstrates that his thought provoking theory still remains no more than an electromagnetic analogy. Sciama died in the following year having left a lasting legacy by the nurturing of a large number of cosmologists at Cambridge and Oxford and beyond who are now at the public forefront of their field. From 1980 to 1984 Sciama also served as president of the International Society of General Relativity and Gravitation.

Chapter 9 References

[9.1]   D. W. Sciama, "On the origin of inertia," *Royal Astronomical Society Monthly Notices*, vol. 113, p. 34-42, 1953.

[9.2]   P. Graneau, N. Graneau, *Newtonian electrodynamics*. Singapore: World Scientific, 1996.

[9.3]   J. C. Maxwell, *A Treatise on Electricity and Magnetism (1873)*. New York, N.Y.: Dover, 1954.

[9.4]   T. E. Phipps, "Should Mach's principle be taken seriously?," *Speculations in Science & Technology*, vol. 1, p. 499, 1978.

[9.5]   Y. Aharonov, D. Bohm, "Significance of electromagnetic potentials in quantum theory," *Physical Review*, vol. 115, p. 485, 1959.

[9.6]   S. Olariu, I. I. Popescu, "The quantum effects of electromagnetic fluxes," *Rev.Mod.Phys*, vol. 57, p. 339-436, 1985.

[9.7]   R. Wilson, *Astronomy through the ages*. London: Taylor & Francis, 1997.

[9.8]   P. Graneau, N. Graneau, *Newton versus Einstein*. New York: Carlton Press, 1993.

[9.9]   P. Graneau, N. Graneau, "Machian Inertia and the Isotropic Universe," *General Relativity & Gravitation*, vol. 35(5), p. 751-770, 2003.

[9.10] A. A. Penzias, R. W. Wilson, "A measurement of excess antenna temperature at 4080 Mc/s," *Astrophysical Journal*, vol. 142, p. 419-421, 1965.

Chapter 10

Retarded Action at a Distance

A Short-Lived Misnomer

Dennis Sciama's valiant effort of trying to merge Mach's principle with Einstein's general relativity may not have proved to be successful, but during the second half of the twentieth century, it was often quoted as inspiration for the continuing search for a realistic explanation of the force of inertia. Several of these attempts utilized Mach's principle with particle interactions involving the distant universe and at the same time studiously avoided Einstein's field theory.

An electrical engineer from the Massachusetts Institute of Technology, Parry Moon, led the charge against general relativity. Together with his former research student and later his wife, the mathematician Domina Spencer, he formulated the first Machian particle interaction theory that aimed to account for inertia. [10.1] In addition, Moon and Spencer proposed an explanation of why the universe could be expanding. To obtain their goals, they modified Newton's law of gravitation in two ways. They multiplied the force of attraction by a number which gradually changed from +1 to −1 as the distance between the interacting bodies increased. When this factor became negative at large separations, the normal gravitational force switched to repulsion, leading to a novel mechanism for driving the apparent expansion of the universe.

The changeover from attraction to repulsion had been postulated in an earlier century, but the real novelty in the Moon and Spencer theory was the abandonment of the simultaneity of distant interactions, while still dismissing the concept that energy or anything else traveled between the interacting particles. This type of force connection has become known as "retarded action at a distance". In essence the retarded forces made this theory numerically equivalent to Einstein's relativity. However Moon and Spencer felt inclined to insist that it was still an action at a distance model in order to comply philosophically with Mach's principle.

This novel concept stood in contrast to Newtonian gravitation, in which the instantaneous action at a distance force between two particles acts mutually and simultaneously and always results in attraction between the particles. Employing a different mechanism, Newton attributed the force of inertia, resisting the acceleration of a body, to contact with absolute space. Since this space was everywhere, the force of inertia did not have to travel from one place to another. It acted instantaneously and locally wherever the acceleration of matter took place. Newton's inertia was therefore a contact action theory, albeit without the reaction force on matter normally required by his own famous third law. Hence no time retardation was involved in either his force of inertia or universal gravitation.

However, things have been conceptually very different in all speculations which have concerned the transmission of light and electromagnetic radiation. The words, "radiation", "transmission" "detection" and others like them all imply a bias that humans have assumed from at least the time of the ancient Greek philosophers. In ancient times, light was assumed to be a God given substance and the casting of shadows seemed to imply that this material traveled in straight lines. The ability to manipulate mirrors to cast light in desired directions further added to the strong conviction that light is a physical entity that travels between a source and a detector. In the last few hundred years, increasingly accurate laboratory equipment has allowed us to measure a time delay between a cause and effect relating a light source and a detector. Since this delay appears to be directly related to the distance of separation, it seems to support the model in which light is a substance that travels at a fixed speed. This mental picture is now universally accepted by all modern textbooks. To cloud the issue slightly, the theory of quantum mechanics is unclear about what it is that is actually moving. Light is now usually described as sometimes a particle and sometimes a wave, but never both at the same time. Both of these contradictory models have their roots in the science of the 17th century.

Scientists at that time spoke of two competing theories of light. Newton [10.2] preferred the corpuscular theory in which small particles of light traveled along straight lines at finite velocity. This mechanism has been resurrected in the twentieth century through the notion of photons which supposedly transport light and all electromagnetic radiation over astronomical distances as well as the short hops of laboratory experiments.

The second theory of light was wave propagation in an ether. Its foremost exponent was the Dutch scientist Christiaan Huygens [10.3]. It was built on an analogy with sound waves which obey the principle that disturbances always travel at finite velocity which depends on the physical properties of the medium through which they pass. The first numerical estimate of the velocity of light was provided by Olaf Roemer (1644-1710) [10.4]. In 1676, he noticed that the timing between eclipses of one of Jupiter's moons was related to the Earth-Jupiter distance. A retardation occurred as this distance increased. It seemed to be caused by a finite and constant light propagation velocity, the magnitude of which could be calculated from astronomical observations. This result however was not accepted for over 50 years until a similar velocity of light was measured by the English astronomer, James Bradley, in 1727 using an entirely different technique.

The present generation of physicists believes the propagation of light at approximately 3×10^8 m/s was firmly established by Maxwell's equations. His famous research aimed to investigate and aid the unification of the new sciences of electricity and magnetism. His mathematics revealed a distance related delay which was numerically related to the material properties of electrical insulators. This prediction did not surprise Maxwell because it agreed quite well with Roemer's measured optical delay. Maxwell was, of course, aware of the competing corpuscular and wave models of light that had been debated throughout the two centuries before him. He also recognized the efforts made by his contemporaries on the continent who were advancing the ideas of both instantaneous and retarded action at a distance. So important was this subject to Maxwell that he returned to it on the closing pages of his treatise [10.5]. There he wrote:

"Now we are unable to conceive of propagation in time, except either as the flight of a material substance through space, or as the propagation of a condition of motion or stress in a medium already existing in space. In the theory of (Carl) Neumann, the mathematical conception called potential, which we are unable to conceive as a material substance, is supposed to be projected from one particle to another, in a manner which is quite independent of a medium, and which as Neumann has himself pointed out, is extremely different from that of the propagation of light. In the theories of Riemann and Betti it would appear that the action is supposed to be propagated in a manner somewhat more similar to that of light"

In this quotation, Maxwell describes Neumann as one of the prime exponents of instantaneous action at a distance theory. He also draws attention to the group which included Riemann and Betti as well as other famous mathematicians such as Gauss and Liénard, who were uncomfortable with simultaneous interactions for they felt that such a mechanism was incompatible with the delayed behaviour of optical effects. Their retarded action at a distance theories were however never widely disseminated and made little impact in the overall development of the subject. Nevertheless, these theories ascribed great significance to what they referred to as a retardation constant, c.

It is worth some attention at this stage to appreciate that the constant, c, now referred to exclusively as the velocity of light, has had an extraordinary history in the development of physics. At different times, it has been used to support both theories of action at a distance as well as field theory. The originators of the letter, c, which stood for nothing more meaningful than "constant" were two German physicists at the University of Göttingen. In 1856, Wilhelm Weber and his colleague Rudolph Kohlrausch performed an experiment to compare the currently prevailing force laws of electrostatics and electromagnetics [10.6]. These two laws, Coulomb's and Ampère's respectively, were both based on the instantaneous action at a distance principle. Weber had been trying to create a unifying formula for electrical forces which had to agree with these two constituent laws. His first theoretical obstacle occurred because the two laws had apparently incompatible units for measuring electricity. While Coulomb's law described the force between two individual charges, Ampère's force of attraction or repulsion existed between current elements which were small lengths of wire which were passing electrical current. At this time, there was no experimental understanding of what occurred inside a wire that was passing current. Weber and his close colleague G. T. Fechner proposed that charges were moving inside a current carrying conductor. Some of their hypothesis has certainly turned out to be correct, but nevertheless Weber had no theoretical way of discovering how many Coulombs of charge had to be traveling through a length of wire in a given amount of time to be equivalent to one Ampèrian current element.

Needing this quantity in order to create his unifying force formula, Weber and his colleague Kohlrausch charged up a Leyden jar with a known capacitance and voltage and thus worked out the stored charge in electrostatic units. They then discharged the jar into a coil of wire and measured the force that the transient current impressed on a magnet, thus measuring the charge in electromagnetic units. They found the ratio of electrostatic charge to electromagnetic charge to be 3.1×10^8 m/s, and this was precisely the constant that Weber required to complete his unified force formula. This was the first measurement of c and clearly showed that it had nothing to do with radiation between separated bodies.

The coincidence that Weber's dimensional constant, c, was equal to the measured speed of light as found by Roemer, Bradley and a growing list of optics experts, was taken up by Gustav Kirchoff, who was at that time developing what we call today "circuit theory". His development of the now well used electrical concepts of inductance, capacitance and resistance stemmed from the electrostatic and electrodynamic theories of Ampère, Neumann, Weber and Coulomb. In 1857, using these principles, he was the first to derive the velocity of voltage and current disturbances down a transmission line [10.7]. Such an electrical circuit comprises two parallel conductors with a source of electrical signal at one end. Kirchoff found that signals travel down the line as if they had a high velocity of the same order of magnitude but always slightly slower than c. However since he was only working with theories of instantaneous action at a distance forces, he never conjectured that anything was actually traveling at this velocity. In fact, his discovery would be equally well presented as the discovery of a delay which described the time taken for a signal to be detected at one point of a transmission line after being recorded at another a known distance away. This delay would have the value of at least $1/c$ or 3.3 ns/m. (1 ns = 1/1,000,000,000 s)

It is interesting to speculate what might have occurred if Weber had described his constant as a distance related delay rather than a velocity. It may have led to novel directions in theoretical physics which have never entered the mainstream. However with c treated as a velocity, it was inevitable that there would be a headlong search for the substance that travels at this speed. It was James Clerk Maxwell, nearly 10 years after Kirchoff's transmission line discovery, who married electromagnetism to the classical wave theories and made the bold interpretation that it was light that traveled at c relative to the ether through which it passed.

There followed a period of intense political rivalry between physicists in the United Kingdom who favored Maxwell's new field theory and the continental European school of action at a distance. While Maxwell was fair in his recognition of his rival scientists, his followers, most notably Oliver Heaviside, George Francis Fitzgerald and Oliver Lodge, who became known as the "Maxwellians" were overtly determined to write the action at a distance theories out of the textbooks.

In 1893 Heinrich Hertz discovered that circuits with oscillating currents produced currents in other metallic loops in other parts of his laboratory [10.8]. He found that there were places in the lab where this pick up was enhanced and other locations where no pickup occurred. The Maxwellians were quick to convince Hertz and the rest of the scientific community that this behavior was easily explained by the passage of radio waves throughout the room. They drew the analogy that light presumably behaved in an identical way to acoustic waves, however with much shorter wavelengths. Historically, this marked the end of the era during which action at a distance theories were seriously considered. However, it was never shown that Hertz's experiments could not be explained by instantaneous action at a distance without the mediation of electromagnetic fields. It is a clear example of the fact that physicists are not immune from allowing fashion to affect their decision making.

Given the lack of modern digital computers in the late 19th century, it is not surprising that physicists pursued a field theory which was soluble by analytical mathematical equations. The solution of problems involving electromagnetic radiation with action at a distance theories could only recently be attempted with computers. Therefore, it is quite understandable that physicists decided to solely pursue field theory at the end of the 19th century. However, it might have been helpful if they had spelt out more clearly that no experimental facts had ever contradicted a possible action at a distance interpretation and that this alternative interpretation would have to be left as a possibility until better computation methods were available. Maybe they simply could not have foreseen the development of computers which would be capable of finite element analysis. The consequence is that there is still the possibility that the effects that we now ascribe to electromagnetic radiation may be explicable by a theory in which nothing travels.

The original arguments between Newton and Huygens about the nature of light remain with us and have resurfaced in the modern wave-particle duality of quantum mechanics. In addition to classical field theory proposed by Maxwell as well as the action at a distance theories of the 19th century German school to which he referred, a third method of light propagation was later proposed by Einstein and has been universally adopted. This concerns the flight of 'lumps of energy' better known as photons which do not require a medium through which to travel. However, the photon theory still needs an additional part to explain experiments that appear to support the wave theories such as refraction and diffraction.

Today it appears that the long-standing wave-particle dual nature of light can be resolved by giving up the notion that light is a substance that travels. Prominent in the body of instantaneous action at a distance theory is Neumann's law of induction, a cornerstone of the old

Newtonian electrodynamics [10.9] which could quite naturally form a basis for this new physics. The first step in this direction was taken by Burniston Brown in the early 1960s with a paper entitled "A new treatment of diffraction" [10.10]

Guy Burniston-Brown was a reader of physics at University College, a part of the University of London in England. As will be seen in the next chapter, London was a hot bed at the time, not just of pop music and swinging fashion, but also of radical new challenges to accepted physics. There was a brief period where one was not censured for questioning conventional Einsteinian relativistic physics. Healthy and heated debate in the journals raged over issues such as the lack of physical evidence of time dilation and length contraction, which were cornerstones of special relativity. Burniston Brown took a very strong stand against making unnecessary hypotheses and consequently felt very strongly that no assumptions should be made with regard to the nature of light. In his highly unconventional approach to optics, Burniston Brown proposed that electromagnetically excited atoms can be treated as oscillators just as Max Planck and other well known physicists had done so before him. Metallic atoms make particularly good oscillators but the atoms of dielectric materials will also oscillate. A property of the atomic oscillator is that, if one is made to vibrate at a certain frequency, other atoms within its sphere of influence will tend to resonate with it.

Burniston-Brown considered a forced oscillator, which will be denoted by O, and a screen consisting of resonating secondary oscillators S. Then he explained [10.10]:

"The source (O) produces oscillatory forces which activate the oscillators in the screen (S). The effect of these oscillators (on each other) is to produce, at a certain distance inward from the front surface (of the screen), a force equal and opposite to that due to the source (O) at every instant, so that the oscillations there, and further in, are not subject to any force and remain inactivated (in darkness). Experiment shows that the layer necessary to achieve this effect is, in the case of metals, extremely thin, amounting to a few wavelengths only. In the case of dielectrics used for screens, the thickness of the layer is very small compared with the distance between the screen and the point of observation."

At this juncture Burniston-Brown should have added that the atomic oscillators of a transparent dielectric, as for example glass, respond hardly at all to the oscillations of O. A detector atom, D, located beyond the transparent screen then interacts with O and is not in darkness. This theory therefore explains the blocking of light with an opaque screen which is often taken as proof that light is 'something' which travels from the source to the detector. In fact it can equally well be explained by an action at a distance theory such as the one outlined above.

Burniston-Brown's light diffraction theory of atomic oscillators is mathematically similar to mutual electromagnetic induction between atomic current elements. We have fully described and extended the use of Neumann's law of induction to the interaction of individual oscillating Ampèrian elements in our book *Newtonian Electrodynamics* [10.9]. For a sinusoidal oscillation of the direction of an Ampèrian current element, which, as we today hypothesize, is a single conductor atom, the mathematical treatment reveals precisely the screening action described by Burniston-Brown. The puzzling aspect of Burniston-Brown's paper is his insistence on retarded action at a distance, whereas Neumann's law of mutual induction is based on instantaneous action at a distance. Burniston Brown suggested that his interactions were both instantaneous and retarded and proposed a

"Law of Retarded Action which states that simultaneity is an important feature of interaction between moving bodies, determining the magnitude and direction of the retarded force. … When retardation is considered, it is the force that must be retarded, not a mathematical term, the potential."

In 1963, at about the time Burniston-Brown developed his light diffraction theory, one of the authors (PG) applied Neumann's law of induction to electrodynamic phenomena in extended metallic conductors [10.11]. This law describes electromotive forces that occur between current elements when Neumann's mutual potential (stored energy) between them changes in time as a result of external influences. The finding was the same as Burniston-Brown's: multiple successive inductive interactions between closed currents and metal filaments - all based on instantaneous action at a distance - will produce phase shifts in the induced current pattern which mimic the progression of electromagnetic waves in the metallic medium. It is very likely, therefore, that instantaneous and remote interactions of metallic atoms can produce effects which may equally well be envisioned as transport of electromagnetic energy at finite velocity.

Burniston-Brown went on to apply his notions of retarded action at a distance that he had formulated for his treatment of light and applied them to a theory which predicted the origin of the force of inertia [10.12]. However he gave no justification of why laws that relate to forces between charges should also apply to forces between masses. Nevertheless, he was not the first to make this ad hoc analogy, a tradition that was started by several astronomers in the 1870's [10.6]. Burniston-Brown's ideas were mathematically complicated. His reliance on delays related to c, even during gravitational interactions, ensured that his results predicted the accepted outcome of all of the supposed tests of Einstein's relativity theory. For instance Burniston Brown's model predicted that particles could not be accelerated beyond c because their inertia would become infinite as predicted by special relativity. He also showed that his theory could explain the anomalous precession of the perihelion of the planet Mercury. 40 years previously, this had been hailed as proof of Einstein's theory of general relativity. However, Burniston Brown's retarded action at a distance theory of inertia as well as his theory of light, fell between two stones because they represented an unacceptable philosophical compromise. He tried to retain the Newtonian dislike of unnecessary hypotheses such as electromagnetic field energy by hanging on to the action at a distance concept. However, in an attempt to easily comply with experiments hailed to be a validation of Einsteinian relativity, he absorbed Einstein's assumptions about the speed of light into his theory. However, he phrased them in non-field theory language. Where Einstein discussed the speed of light, Burniston Brown described a distance related delay in the force between two particles.

More recently one of us, NG, has developed a new theory of light [10.13] that also has its origins in the laws of instantaneous action at a distance electrodynamics. It involves the application of Ampère's force law to large systems of current elements. This famous equation, once described by Maxwell as "the cardinal law of electrodynamics", is directly derived from Neumann's potential formula and produces two types of motion as a consequence of the energy stored between two current elements. It determines the strength of the mutual attraction or repulsion, but less widely discussed is the prediction that each element attempts to also change the direction of the current flowing in the other element. If a current element is depicted by a vector whose length describes the current strength and whose orientation defines the direction of the current, then these vectors can be seen to change each other's direction as a consequence of Ampère's force law.

The structure of the model is based on the same principle as the Burniston Brown theory. It hypothesizes that a group of current elements, forced to change their direction or oscillate in a coherent manner, can simulate an antenna or other oscillating source. If we imagine several other groups of current elements separated by predefined distances, then we can model how all of these elements interact with each other with the aid of finite element analysis using a personal computer. Three different types of object were modeled, namely a source, a reflector and a detector. At the beginning of the calculation, all of the elements in the source have the same current strength and direction and are programmed to rotate with the same angular frequency. The elements in the reflector and detector start the calculation with randomly distributed directions and rotational velocities. The calculation rules are simply that there is direct interaction between the source and the reflector, but no interaction between the source and detector, based on the supposition that they are separated from each other by a screen. However, there is direct interaction between the reflector and the detector. When the program starts it creates a sequence of time slices and calculates the directions and rotational velocities for every element. At each time slice, the elements all interact and produce a force on each other that accelerates their rotational motion. The program picks up coherent oscillatory motion in the reflector and detector which can be plotted with respect to time This coherent motion can be interpreted as an alternating electric signal or the detection of a certain frequency.

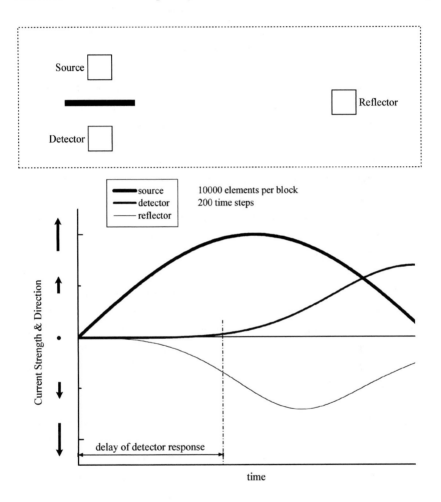

Figure 10.1 : A schematic description of the source, reflector and detector in the Ampère light calculation. The plotted results demonstrate the delay of response at the reflector and detector.

This simple instantaneous action at a distance modeling produces, in these three bodies, the most fundamental properties normally ascribed to electromagnetic radiation. In figure 10.1, the signal at the three locations is plotted. There is a delay of the signal observed at the reflector and a further delay before the elements at the detector start moving coherently.

Even though the mutual forces are instantaneous, the delays appear due to the angular inertia of each element which limits its rotational acceleration. These elements are composed of matter which has mass and therefore they must have an inertia to rotation as well as to linear motion. This concept is never considered in the conventional model of radiation in which it is tacitly assumed that an oscillator in a detector acquires its full rotational velocity at the instant of the arrival of a photon. The conventional photon model completely disregards the effect of inertia. While Burniston-Brown used mysterious delays to explain the existence of the force of inertia, in contrast, the model just presented here uses inertia to explain the existence of the observable delays in electromagnetic signals.

Returning to the Ampère force model, like Newtonian gravitation, the force between elements decreases as the inverse square of their separation. The larger the distance between elements, the weaker the force of the interaction, and thus the lower the angular acceleration of the elements. The computer model which involves millions of element interactions and can run on a PC for days, yields the result that there is the appearance of a delay which depends linearly on distance of separation.

This apparent delay is the piece of the jigsaw that Burniston-Brown and his retarded action at a distance colleagues did not possess. They did not foresee that a theory based on instantaneous mutual forces could predict observable delays. As a result they were forced into a philosophical purgatory.

Figure 10.2 reveals three further important qualities of light predicted by the Ampère force model, which are (a) a precise frequency pickup in the detector (b) a 180° phase reversal on reflection and (c) a decrease in amplitude with distance. These results provide a promising prospect for a purely instantaneous action at a distance theory of light.

The vector model of the rotating element utilised in the Ampère model is identical to the central entity in the theory of Quantum Electrodynamics (QED), called the probability amplitude [10.14]. The originator of this universally accepted theory of light and matter is the eminent physicist Richard Feynman. He simply postulated the behaviour of small abstract arrows (probability amplitudes) that are attached to matter, but used the model of travelling photons to explain their relationships. A huge effort among thousands of physicists has now shown that this theory applies to all of physics except gravitation and nuclear physics. The development of the Ampère force light model demonstrates that the travelling light model is no longer required to explain any electrodynamic effects.

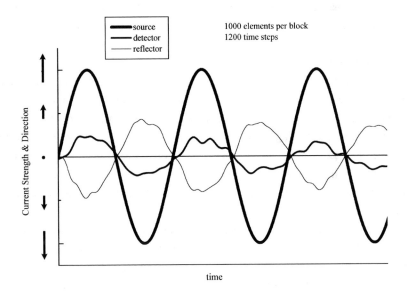

Figure 10.2 : Demonstration of frequency pickup at the reflector and detector.

Further aspects of the instantaneous action at a distance model show that it could be the mutual simultaneous interactions of all of the elements in the universe that conspire to give us the universal constant, c, the speed of light. The value of this constant is hypothesized to be due to the particular distribution of matter in the universe. This means it could be a local constant with different values in different parts of the universe. Perhaps, c could vary in time if indeed the universe is evolving and its structure and mass distribution is changing. This conjecture, involving simultaneous electromagnetic interactions between all of the atoms in the universe, mimics the Machian interactions that we have been describing throughout the book in which inertia is also a property which is defined by the entire universal matter distribution.

In summary, there are still two divergent theories that describe the effects commonly called electromagnetic radiation. One is the conventional theory of a traveling substance called light. It is still troubled by paradoxes, most notably the wave particle duality. However, more problematic is that the light itself can never be directly detected and subjected to objective study. All that can be analyzed are the relationships between sources and detectors.

In contrast, action at a distance theories relate precisely to the known facts and make no assumptions regarding an unobservable traveling substance. We would hope that most people would still agree with Newton's guidance in that the best way to do science is to make the least hypotheses. Therefore the action at a distance models must always be pursued if at all possible. Clearly, some theoreticians never imagined that an instantaneous action at a distance model would be able to account for the well known delays found in radiation experiments. Thus the notion of retarded action at a distance was invented. However, we can now see that this term is really a misnomer for these theories predict delayed forces even though the interactions are simultaneous. As a result of this contradiction, this concept has historically dropped away from present research agendas.

However the authors of this book have been developing two related models involving instantaneous action at a distance interactions which show that it is possible to explain the delays that pertain to electromagnetic "radiation" without need for hypotheses regarding traveling light. This returns the concept of instantaneous action at a distance to full acceptability. Consequently the philosophically inconsistent retarded action at a distance inertia theories of Moon and Spencer and later of Burniston-Brown are not relevant to the origin of inertial forces. Further we are no longer bound to Einsteinian theories which require that all interactions between separated matter be restricted by the speed of light. To tie this whole paradigm together, the final chapter of this book proposes a purely instantaneous action at a distance theory which can successfully produce a consistent explanation of inertia in the light of Mach's principle.

Chapter 10 References

[10.1]   P. H. Moon, D. E. Spencer, "Mach's Principle," *Philosophy of Science*, vol. 26, p. 125-134, 1959.

[10.2]   I. Newton, *Optiks (1730)*. New York: Dover, 1952.

[10.3]   C. Huygens, *Traite de la lumiere*. Leiden, 1690.

[10.4]   E. Whittaker, *A History of the Theories of Aether and Electricity (1951)*. New York, N.Y.: Dover, 1989.

[10.5]   J. C. Maxwell, *A Treatise on Electricity and Magnetism (1873)*. New York, N.Y.: Dover, 1954.

[10.6]   A. K. T. Assis, *Weber's Electrodynamics*. Dordrecht, The Netherlands: Kluwer Academic, 1994.

[10.7]   P. Graneau, A. K. T. Assis, "Kirchoff on the motion of electricity in conductors," *Apeiron*, vol. 19, p. 19-25, 1994.

[10.8]   H. Hertz, *Electric Waves (1893)*. New York, N.Y.: Dover, 1962.

[10.9]   P. Graneau, N. Graneau, *Newtonian electrodynamics*. Singapore: World Scientific, 1996.

[10.10] G. Burniston-Brown, "A new treatment of diffraction," *Contemporary Physics*, vol. 5, p. 15-27, 1963.

[10.11] P. Graneau, "Steady-state electrodynamics of a cylindrical body in axial motion," *Journal of Electronics and Control*, vol. 14, p. 459, 1963.

[10.12] G. Burniston-Brown, *Retarded action at a distance*. Luton: Cortney, 1982.

[10.13] N. Graneau, "Have you seen the light ?," in *Instantaneous action at a distance in modern physics: "pro" and "contra"*, A. E. Chubykalo, V. Pope, and R. Smirnov-Rueda, Eds. Commack, NY: Nova Science, 1999

[10.14] R. Feynmann, *QED, The Strange Theory of Light and Matter*. Princeton: Princeton University Press, 1985.

Chapter 11

Clock Confusion in the 20th Century

The Connection Between Inertia and Timekeeping

The theory of General Relativity was published in 1915. A casual reader of the history of science may gain the impression that twenty years later Einstein's work was fully integrated into the teaching and thinking of physics. Teaching – yes; thinking – no. The backbone of the profession, that is experimental physicists in universities, government organizations, and industry, thought it hardly worthwhile to grapple with the complex mathematics of relativity, because the theory was of little or no help to the solving of their everyday problems. It seemed to be more of a philosophical adornment, a conversation piece with which to boast about one's intellectual prowess.

Cosmologists were an exception to this trend. They had the luxury of being able to let their imagination roam without accountability to controllable laboratory experiments. In 1919, Einstein became an overnight international celebrity when it was announced that photographs taken during a full solar eclipse revealed the deflection of light from stars near the sun by just the amount predicted by General Relativity. However within a decade, it had become clear that the Cambridge astronomer Arthur Eddington, who had performed the measurements, had ignored 85% of his data, including stars that were apparently shifted in the wrong direction [11.1]. Unfortunately. this misuse of data was not only held up to promote Einstein's theory, but also to discredit Newtonian gravity. This was even more surprising since Newton had never described a relationship between light and gravity.

It seems that nothing could prevent the rise of Einstein's reputation and his theories and personality entered the accepted folk lore of all nations. Claims are still made that only General Relativity can explain certain astronomical observations, but this cannot be true when rival theories are not being seriously considered. Nevertheless, in the middle of the 20th century, the intellectual elevation of Einstein's theories had still made very little impression on the armies of scientists, engineers and technicians who invented, developed, and produced the machines of our modern world. Instead, they were completely satisfied with the Newtonian mechanics with which they were utterly familiar and which is so much easier to apply.

At the same time the Einstein model of the universe had become the accepted way of describing nature. Newtonian physics was demoted to a useful approximation and any thought that the distant universe had an instantaneous effect on earth was entirely ruled out. This is why Mach's principle and the force of inertia are no longer discussed. Physics undergraduates are now presented with a large body of experiments which they are told demonstrate that Einstein's theories of relativity are proved beyond all doubt. Most of these experiments attempt to investigate Einstein's prediction of the dilation of time. This involves the observation of some very clever clocks in highly unusual situations. The understanding of these tests is the primary subject of this chapter.

As well as enthusiastic supporters of Einstein's ideas, there have always been capable and respectable physicists around who spoke up against them. They have published many papers and books in spite of fierce opposition from other professors and editors of major physics journals. The scientists and journalists, who form the visible core of the physics profession, are apt to tell such dissident authors that the overwhelming majority of their peers are completely convinced that Einstein's world view is unshakeable. Consequently any criticism represents bad science. How can this attitude rest easily alongside Einstein's letters which he wrote in the middle of the twentieth century to his old friends in Switzerland towards the end of his life [11.2], in which he claimed to be unsure of the validity of his theories?

Einstein's failure to find a unified theory of electromagnetism and gravity, his protracted disagreement with Niels Bohr about the probabilistic nature of quantum mechanics and not least the growing realization that Newtonian action at a distance was no spookier than an energy filled spacetime vacuum field, were major factors behind his self doubt. Since the early 1970's, experiments have been performed that demonstrate that quantum detectors interact with each other instantaneously irrespective of their distance of separation [11.3]. Einstein and later physicists in the 1960's such as the Irishman John Bell working at CERN had deduced that if quantum mechanics were true, it required such non-local connections between all of the objects in the universe. There is now a growing body of empirical evidence which goes under the banner of "quantum entanglement" which demonstrates instantaneous non-local interactions. These results have been steadily eroding the foundation on which the field and Einstein's local action relativity theories stand. In 1949 Einstein predicted that none of his concepts were likely to survive and he was not even on the right track to penetrate the secrets of nature. Therefore in the latter half of the twentieth century, it seemed the die was cast for a major paradigm change in physics. But this revolution has still not occurred, presumably as a result of a lack of consensus on an alternative outlook.

London Rebels

Without reference to Einstein's disillusionment, three prominent English physicists based in London launched a strong campaign against Einstein's theory of special relativity (SR) during the 1950-70's. Their criticism was directed on that part of Einstein's theorizing which addresses what happens to objects in "inertial" motion relative to "inertial" observers. The adjective "inertial" must be clearly understood to describe a body which is not being subjected to an external force. The consequence is that this object is not accelerating nor feeling the force of inertia. The names of the rebels were Guy Burniston-Brown, Louis Essen, and Herbert Dingle. Burniston-Brown produced several papers questioning the philosophical integrity of SR [11.4] as well as developing a Machian theory of inertia, based on the retarded action at a distance model which was discussed in chapter 10.

Louis Essen was a British government experimentalist who distinguished himself in the National Physical Laboratory (NPL) as a pioneer of atomic clocks. One suspects that all clock makers are not entirely happy with Einstein's concept of time dilation and Essen became their spokesman. His cesium clocks were highly valued developments to both sides of the cold war. He was the first foreign recipient of the Popov medal, the top Russian physics prize in 1959 as well as receiving a British OBE in the same year. Despite subsequently becoming a Fellow of the Royal Society, he was shunned for his criticism of SR. His primary concern was that Einstein had made a fundamental error with his units in the assumptions of the SR theory. He argued that Einstein had assumed that the speed of light was a fundamental constant, and his formulae constantly adjusted the unit of time to keep it so. To Essen, this was illogical and defied the basic understanding of the process of physical measurement. Essen sacrificed the good will of many of his colleagues to take this radical stance and even now the NPL web page, which highlights many of Essen's discoveries, admits that he was actively encouraged by his employers and the government to suppress his dissident views just prior to his retirement. In 1978 he published an article entitled, *Relativity and Time Signals*, in the journal, Wireless World, [11.5] in which he wrote:

"No one has attempted to refute my arguments, but I was warned that if I persisted I was likely to spoil my career prospects. ... the continued acceptance and teaching of relativity hinders the development of a rational extension of electromagnetic theory."

Best remembered amongst the three Londoners was Herbert Dingle who was a professor of physics at Imperial College and later held the Chair of History and Philosophy of Science at University College, both part of London University. He also rose to the position of President of the Royal Astronomical Society.

Dingle had lectured on relativity at the University of London and published books with titles like *Relativity for All (1922), The Special Theory of Relativity (1940),* and *Mechanical Physics (1941)*. He was an excellent communicator and when he adopted an anti-SR stance in the latter part of his career he caused great trouble to the editors of respectable physics journals, including 'The Proceedings of the Royal Society', 'The Philosophical Magazine', and 'Nature'. His early battles are recounted in *A Threefold Cord: Philosophy, Science, Religion* [11.6]. This book recounts a dialogue with Viscount Samuel, the distinguished British liberal politician who sat in the Asquith Cabinet at the outbreak of the first world war. After the war he became the first High Commissioner of Palestine. At the age of fifty Samuel turned away from politics and devoted his time to philosophy. His principal objective was to find some common ground between philosophy, science, and religion. This led to his conception of a *Threefold Cord,* a book which contains no mathematics and is easily understood by all who are interested in the laws of nature.

Samuel was a firm believer in some form of ether and Dingle could not shake Samuel's faith in this abstract concept. Nevertheless, when Dingle discussed time dilation and the twin paradox, which states that a space traveler ages slower than his twin brother on earth, Samuel was ready to admit:

> ".... but I feel that any theory which is in such flagrant contradiction with common sense would need much more powerful arguments before it would be likely to command any measure of general support."

To this Dingle replied:

> "Your reaction to my account of this controversy is that to be expected of any intelligent person whose reasoning power has not been destroyed or paralyzed by over-indulgence in symbol manipulation: it is that of incredulity."

This shows that Dingle was very much aware how, in the twentieth century, physics had become dominated by mathematics (symbol manipulation) to the detriment of physical models which were based on observational evidence (common sense). In the end, physics must be expressible in words if it is ever to become comprehensible.

Unfortunately, students are now regularly taught that physics is written solely in the language of mathematics and often defies "common sense" and they had simply better get used to it. Dingle along with the authors of this present book believe that intelligible physics must be at least the goal if not the outcome of any theory. Dingle demonstrated the dangers of over reliance on mathematical theory with his proof of a massive internal contradiction in Einstein's theory of SR.

The word "relativity" has traditionally meant that if two bodies move relative to each other, it is not possible, by experiment or otherwise, to claim that one of them is moving more than the other. Relativity was obvious to scientists of the eighteenth and nineteenth centuries. They saw little need to discuss the subject. This natural relativity is often called Galilean relativity as the Italian astronomer had apparently demonstrated that from below decks on a calm sea, it was impossible to measure the steady speed of a ship without looking out of a port hole.

Einstein defined the principle of relativity in a more complicated way. In his first paper on SR [11.7], and later translated into English, [11.8] he said:

"The laws by which the states of physical systems undergo change are not affected, whether these changes of state be referred to the one or the other of two systems of coordinates in uniform translatory motion."

For example, if a train moves through a station with a constant velocity, any event inside the train or on the platform can be analyzed by any observer either on board or on the ground using the same laws of physics. It becomes purely a matter of tradition who is considered to be moving and who is stationary.

However, as Dingle explained with great clarity, the symmetry of relative motion breaks down if the predictions of SR are correct. He referred to the clock paradox, which is also known as the twin paradox, and has been discussed extensively in both the specialist and popular physics literature throughout the twentieth century. Einstein claimed that SR predicts that if one of the twins goes space traveling and later returns home to earth, he finds himself to be younger than his brother. Somehow in Einstein's world model, space travel involves more motion with respect to the earth than the motion of the earth with respect to the space ship. This is purported to result in asymmetrical aging.

In 1972, ten years after the publication of the *Threefold Cord*, and after several more public dialogues with eminent physicists and philosophers in well respected journals, Dingle had still not achieved his goal of encouraging a general rethinking of Einstein's theory. As a consequence, he published another book, *Science at the Crossroads* [11.9]. It is a book which he conceded he had not wanted to write. However he felt obliged to point out that expensive and dangerous physics experiments were now being undertaken in laboratories all over the world and he was convinced that they were being designed based on an implausible theory. By 1972, after thirteen years of honing his argument down to the simplest possible exposition he wrote:

"It would naturally be supposed that the point at issue, even if less esoteric than it is generally supposed to be, must still be too subtle and profound for the ordinary reader to be expected to understand it. On the contrary, it is of the most extreme simplicity. According to the theory [special relativity], if you have two exactly similar clocks, A and B, and one is moving with respect to the other, they must work at different rates, i.e. one works more slowly than the other. But the theory also requires that you cannot distinguish which clock is the 'moving' one; it is equally true to say that A rests while B moves and that B rests while A moves. The question therefore arises: how does one determine consistently with the theory, which clock works the more slowly? Unless this question is answerable, the theory unavoidably requires that A works more slowly than B and B more slowly than A – which it requires no super-intelligence to see is impossible. Now, clearly, a theory that requires an impossibility cannot be true, and scientific integrity requires, therefore, either that the question just posed shall be answered, or else that the theory shall be acknowledged to be false. But, as

I have said, more than 13 years of continuous effort have failed to produce either response."

In 2005, the 100th anniversary of the birth of the theory of SR and 33 years after Dingle's last plea for logic to prevail, nobody has answered his question and Einstein's theory is still held to be the bedrock of modern physics. This overwhelming level of support for SR is based on several famous experiments which seem to numerically support some of its predictions. If Dingle was correct in his reasoning, this situation can only have arisen if generations of physicists have not been applying the theory to experiments correctly. We show later in this chapter that this is precisely what has occurred.

The Foundations of 20th Century Physics

What is the reason for asymmetrical aging in special relativity ? Or more importantly, why is it absent in Newtonian physics? The answer can be found directly in Einstein's motivation for creating the theory. By the end of the 19th century, the accepted foundations of physical theory rested on two pillars, the Machian reinterpretation of Newtonian physics and the Maxwell-Lorentz theory of electromagnetism. Mach's mechanics is often quoted as Einstein's primary inspiration for his emphasis on the importance of Galilean relativity.

The difficulties that Einstein confronted, arose when it became clear that Maxwell's field equations were not invariant under Galilean transformations. In less technical jargon, this means that according to Maxwell's theory, the strength and nature of the electromagnetic fields surrounding a body depended on its absolute velocity with respect to an ether. This comes about because Maxwell's equation's are built around a constant called c, which he defined to be the absolute velocity of propagation of electromagnetic fields through an ether. It seemed to everyone a natural analogy to the highly successful theory of acoustic waves which travel at a speed which depends entirely on the physical properties of the medium through which they flow. c is of course now universally described as the speed of light. However, to Einstein, the undetectable ether felt very similar to the abhorrent notion of Newtonian absolute space. Mach's rejection of the concept of absolute motion clearly inspired him to seek a way of philosophically rescuing Maxwell's equations. In his theory, he eliminated the ether, and ensured that the physics that one observed did not depend on the absolute velocity of one's steady motion.

Einstein's argument rested on the assumption that the speed of light, c, was a universal constant for all inertial (moving with a steady velocity) observers. As Louis Essen has pointed out, Einstein somehow managed to persuade his peers that his assertion regarding the constancy of the speed of light for all observers was more fundamental than keeping a well understood and consistent unit of time. Remarkably it seems that Einstein's proposal to overthrow all previous concepts of time was greeted with great enthusiasm, whereas it may have been prudent to consider other possibilities before jumping headlong into the destruction of conventional timekeeping. The bold assumptions of SR also directly conflicted with the long established and highly successful theory of Newtonian mechanics.

Einstein's drastic measures were clearly considered an acceptable sacrifice in order to save the Maxwell-Lorentz electrodynamics. It seems that very little attention was paid at the time to any alternative theories of electrodynamics which would not have required a radical distortion of space and time in order to satisfy Galilean relativity. History has somehow forgotten that there was indeed another available theory of electrodynamics based on the action at a distance laws of Andre Marie Ampère, Franz Neumann, Augustin Coulomb, Wilhelm Weber and Gustav Kirchoff. This philosophically distinct approach to the subject was highly praised by Maxwell who actively encouraged his readers to keep an open mind and let future discoveries determine which approach was more accurate. This now forgotten body of understanding has been reviewed in our earlier book, *Newtonian Electrodynamics* [11.10] in which it is demonstrated that the relativistic Maxwell-Lorentz field theory cannot be applied to all situations and a return to an action at a distance Newtonian electrodynamics is urgently required. The acceptance of such a field free theory would have removed the need for the invention of SR and the consequent distortion of the units of space and time. However history tells another story, the glorification of Einstein's imagination.

The Michelson-Morley Myth

The theory of special relativity achieved its goal of making Maxwell's equations invariant for all inertial observers. However, the weakness of Einstein's theory lies in the fact that it was not built empirically upon a body of solid experimental knowledge. Instead Einstein based his model on the unsubstantiated assumption that the speed of light is constant for all unaccelerated observers. In his seminal paper in which he presented the theory of relativity in 1905 [11.7], Einstein provided no references at all and certainly gave no clue regarding what information he had used to justify his assumption. Even though Einstein later claimed that he was unaware of it at the time, his colleagues and followers soon started quoting a now famous experiment by the American physicists, Albert Michelson and Edward Morley, as the evidence that confirms Einstein's assumption.

The elaborate and expensive test that is now universally referred to as the Michelson-Morley experiment was performed at the Case School of Applied Science in Cleveland, Ohio and published in 1887 in the American Journal of Science under the title, *On the Relative Motion of the Earth and the Luminiferous Ether* [11.11]. The scientists were clearly under the influence of Maxwell's electromagnetic ether model and were attempting to prove its existence by measuring the speed of light in two orthogonal directions at the same time, using a device now called a Michelson interferometer. This experiment is analogous to trying to measure the strength of the current in a river by taking two identical swimmers and timing their return trips over 50 meter courses in two directions at right angles to each other. If their times are different, then one can calculate the speed and direction of the water flow. One can also make the clear deduction that identical swimmers can move at different speeds relative to the river bank depending on external conditions such as the current. Instead of swimmers, Michelson and Morley attempted to measure the speed of light in differing directions to determine whether external conditions affected it. The details of the

Michelson-Morley experiment are not as important as the calumnious manner in which the results of this experiment have been represented over the intervening years.

Without exception, all modern undergraduate physics textbooks report that the Michelson-Morley experiment is the most famous and important null result in the history of science. In other words, they use this famous paper to confirm that no differences in the speed of light were found in any direction. This is however not what Michelson and Morley reported. They wrote

> "... the relative velocity of the earth and the ether is probably less than one-sixth the earth's orbital velocity, and certainly less than one-fourth. ...The experiment will therefore be repeated at intervals of three months, and thus all uncertainty will be avoided."

Unfortunately, they never did repeat the experiment at different times of the year, but most importantly they certainly did not report a null result. The measured speeds were simply less than they expected and getting near the limits of the resolution of their equipment. There is certainly no finding in the paper strong enough to justify Einstein's construction of a completely new physical model. However, the null result interpretation clearly became attractive to Einstein and his followers if it meant that Maxwell's theory could be saved.

Similar experiments were performed by Morley and one of his students, Dayton Miller, in the first few years of the 20th century. Like the original experiment, they suffered from insufficient readings to make a solid case. Nevertheless, they consistently revealed evidence of differences in the speed of light. In 1921, two years after the publication of Eddington's eclipse data and the wide acceptance of General Relativity, Dayton Miller was visited by Einstein and they both felt that it was imperative to determine once and for all whether the Michelson interferometer produces a null result, as required by SR. Miller was awarded a lavish research budget and set up the most elaborate interferometer to date. He made measurements with it in Cleveland and at the Mount Wilson observatory in Southern California.

The most thorough set of experiments was performed at Mount Wilson between 1925 and 1926. In this period, he took over 100,000 readings from 6,402 turns of the interferometer. The readings were taken in four batches, separated by three months, to investigate the effect of four epochs of the earth's orbit around the sun. In comparison, the original Michelson-Morley data was taken during a single four day period involving only 36 turns of the device. In addition, Miller had taken the previous two years to perform control experiments which involved subjecting his apparatus to known mechanical and thermal distortions so that such effects could be eliminated from the final experiment. The history and final results of his investigations were finally published in 1933 [11.12].

In order to appreciate the magnitude of Miller's discoveries, it is necessary to understand one astronomical concept, that of the sidereal day. A conventional 24 hour "civil" day is the time required for the sun to reappear the next day at the same east-west longitude. This is the day that we measure on our watches. In this time, however, the earth has completed a small portion of its orbit around the sun and consequently, at midnight, the sphere of background stars that we observe has rotated by a few degrees from the previous midnight. The amount of civil time between the reappearance of the fixed stars in the same location for a given observatory is actually 23 hours 56 minutes and 4 seconds. This is the definition of a sidereal day which is split evenly into sidereal hours, minutes and seconds.

Miller reported in 1925 that after analysing the data from the first three epoch periods :

"the curves for the three epochs were simply averaged and it was found that when plotted in relation to *local civil time*, the curves are in such phase relations that they nearly neutralize each other; the average effect for the three epochs thus plotted is very small and unsystematic. The curves of observation were then plotted with respect to *sidereal time* and a very striking consistency of their principles was shown to exist, not only among the three curves for azimuth and those for magnitude, but, what was more impressive, there was a consistency between the two sets of curves, as though they were related to a common cause. The average of the curves, on sidereal time, showed conclusively that the observed effect is dependent upon sidereal time and is independent of diurnal and seasonal changes of temperature and other terrestrial causes and that it is a cosmical phenomenon."

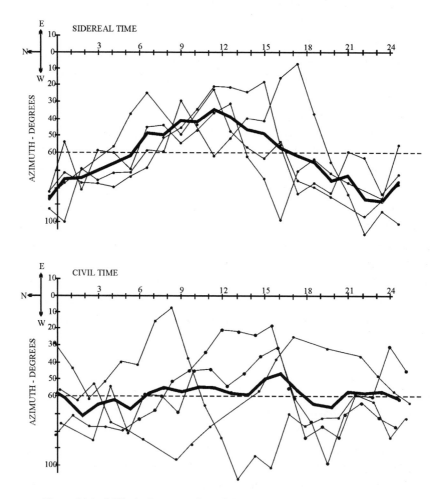

Figure 11.1 : Miller's demonstration of the dependence between his positive effect and sidereal time (from [11.13]). Thick line is the average. The second graph shows no dependence between his results and civil time.

It is quite surprising that after 40 years of this type of experimenting, nobody else had plotted the data against sidereal time. However, only Miller had taken such a large amount of data spread throughout an entire year. Over the course of a few days, there is not much shift between civil and sidereal time, but over the course of three months, the two time bases get out of synchrony by 6 hours. The difference between the averaging with respect to the two time scales is shown in figure 11.1. The "azimuth" simply represents the compass direction at which the experiment produced the most effect at any given time. Clearly the average with respect to sidereal time reveals a true phenomenon which displays a direct interaction with the fixed stars.

When a scientific experiment is being designed, there is always one or more hypotheses under investigation. By the time Miller came to design his equipment and experimental timetable, he was attempting to investigate at least half a dozen differing ether hypotheses that had been proposed by various scientists over the previous forty years to explain the Michelson-Morley results. The most famous of these included a static ether as Michelson had first proposed, or an alternative was an ether that was static in the universe but was locally pushed by the earth as it moved through it. There was of course Einstein's hypothesis that there was no ether at all which went with his prediction that the equipment became shorter in the direction of motion. Miller actually wanted to divorce himself from all preconceived theories and directly discern whether he could measure what he called the "absolute motion of the earth" with respect to the distant stars.

Miller's predecessors had only been concerned with the possible effects on a terrestrial experiment due to a local ether wind. As a result, they failed to appreciate two important aspects of their results. They knew that the orbital speed of the earth around the sun was approximately 30 km/sec, and therefore assumed that the ether wind would be at least this speed or more. When their experiments produced results that were non-zero, but nevertheless lower than their expectations, they assumed that the positive results were erroneous. Miller felt strongly enough that these results were inconclusive and fortunately had the foresight, determination and, most importantly, the funding to make more accurate measurements at four different times of the year. From his vast volume of recorded data, Miller eventually concluded that the only interpretation of his results was that the earth and solar system were moving against the backdrop of distant stars with a velocity in excess of 200 km/sec in a direction toward a star in the constellation Dorado in the Southern Sky. Unfortunately his interpretation of the cause of his findings was inevitably based on his own assumed version of the ether. Thus his final predictions regarding the motion of the solar system are also inconclusive. However Miller's data definitely confirmed that the speed of light is not the same in all directions with respect to the background stars

In 1921, Einstein was very concerned by the preliminary positive results of Miller's experiments. He wrote to his colleague Robert Millikan [11.14]:

"I believe that I have really found the relationship between gravitation and electricity, assuming that the Miller experiments are based on a fundamental error. Otherwise, the whole relativity theory collapses like a house of cards"

As of today, nobody has discovered a "fundamental error" in Miller's results. However in the early 1950's, Miller's successor in the physics department at Case Western Reserve University, Robert Shankland, formed a close relationship with Einstein and undertook to revaluate Miller's data. In a paper [11.15] published in 1955, fourteen years after Miller's death, Shankland's team analysed several of the 24 hour data series and revealed that the results were indeed quite noisy, induced primarily by temperature variations, which naturally occur during any given day. Miller had foreseen this problem and this is why he took such an overwhelming amount of data to try and average out these experimental distortions. At no point in Shankland's lengthy analysis did he take into account that Miller had found a strong dependence on sidereal time as opposed to civil time. The temperature variations to which Shankland paid attention were due to changes between night and day which clearly depend on civil time but over the course of a year have no relation to sidereal time To Miller and his supporters, the correlation with sidereal time proved that the speed of light depends on direction with respect to the fixed stars. This aspect was completely ignored in Shankland's incomplete analysis.

Not surprisingly given the overwhelming support for Einstein's relativistic theories by the mid 1950's, Shankland's dismissive paper has become the celebrated final accepted word on the issue. Unfortunately, since Shankland's investigation, the vast number of data sheets to which he had access in the Case Western Archives have disappeared.

So successful was Shanklands discrediting of Miller's conclusions, that a Michelson-Morley type experiment performed in 1964 [11.16] never even considered to take data over the course of a year so that an effect with respect to sidereal time could be investigated. Following the logic of Michelson, this group only tested the hypothesis that the earth was possibly moving through a static ether. As a result they based their findings on results taken only over the course of a single day. Not surprisingly, they came to the conclusion that there is no discernible effect on the speed of light due to the earth's motion around the sun.

Fortunately, the Miller results have been reinvestigated at least once more, notably by the very eminent French physicist and economist, Professor Maurice Allais. He won the Nobel Prize in economics in 1988 for his work on maximising the efficiency of national economies. However, his professed true love was fundamental science and in 1978 he was awarded the Gold Medal of the National Centre for Scientific Research, the highest honour in French Science, for his contributions. He had a particular interest in gravity and during the 1950's was widely lauded for his discovery of still unexplained gravitational anomalies during eclipses. In the 1990's Allais unearthed the work performed by Miller 70 years earlier and went as far as declaring the current teaching of this subject to be a "cover up". In one of his recent papers on the subject [11.17], he wrote

"The highly significant regularities displayed by Miller's observations do correspond to a very real phenomenon which cannot by any means be attributed to temperature effects. Consequently the light velocity is not invariant to its direction over time. As a result Einstein's special theory of relativity is based on a principle, the invariance of light velocity, which is contradicted by observation data"

It is tragic that during the last 80 years, nobody has successfully performed an experiment like Miller's. As a result, scientists are using Einstein's theory of relativity with complete confidence even when the foundation of its assumptions have been experimentally disproved. Unfortunately the Miller experiments are now either completely ignored or discredited and seen purely as an unsuccessful negative attack on the conventional understanding of modern physics.

In contrast, a positive interpretation of the results of not only Miller, but also his fellow interferometer experimenters, is that terrestrial physics is directly affected by the distant universe or what Mach referred to as the fixed stars. In the 19th century, it was initially quite shocking that Foucault's pendulum appeared to display a connection with the fixed stars. However, eventually this was seen to be an exciting advance on our knowledge of our position in the universe. While perhaps similarly disturbing to field theorists who believe that all of physics is restricted to local phenomena, eventually Miller's results will force us to come to terms with the fact that the speed of light is not constant and also has a connection with the distant cosmos.

The Experimental Tests of Special Relativity

Length Contraction ?

In 1905, Einstein's assumptions were considered to be reasonable and the perceived theoretical benefit was the survival of Maxwell's electromagnetic equations. However, controversies soon arose when the theory was applied to the motions of real material objects. SR makes predictions about physical changes to moving objects. It states that if we take two identical objects, A and B, both at rest with respect to each other, and then accelerate one or both of them to a high velocity, then if we can observe object B from the rest frame of A, it will appear shorter than when the two were at relative rest. This is the famous notion of

Lorentz length contraction which has never been tested experimentally because of technical difficulties.

The Lorentz length contraction leads to a famous theoretical paradox, first described by Paul Ehrenfest within four years of the publication of Einstein's theory. He considered a solid spinning disk in which the periphery can be considered as a chain of very short, virtually straight rods. When the disk is rotating, each of these rods should contract in length, thereby reducing the circumference. In contrast if the spinning disk is described as collection of radial rods, these retain their original length since they are not in radial motion. The Ehrenfest paradox asks how a rotating disk can reduce its circumference while retaining the same radius.

Thomas Phipps, a contemporary American critic of relativity theory has beautifully compiled the many reactions to Eherenfest's dilemma in his book, *Heretical Verities* [11.18]. He describes the first public responses to the Ehrenfest challenge by two mathematicians who published separately, but whose work was later compiled as the Herglotz-Noether theorem. In essence, they predicted that as a result of SR, a rigid disc cannot spin. Physicists at the time were not completely satisfied with this solution since they, like all of us, had observed many rotating disks, and as a result came to the conclusion that there was no such thing as a perfectly rigid body and therefore spin was actually possible. This solution completely ignored the fact that Einstein's theory of SR only applies to perfectly rigid bodies. According to Phipps, this allowed a new concept to enter the world of physics, the "impermissible idealization". With such a notion, it becomes possible to have a theory that can never be tested experimentally. For the majority of physicists and mathematicians who accept Einstein's theory, this apparently causes no concern, but for others typified by Dingle and Essen, it meant that SR is a useless theory.

Mass Variation ?

Another prediction of the special theory of relativity is that as an object gains velocity, it also increases in mass. As the matter approaches the speed of light, its mass is supposed to tend to infinity, thus dramatically increasing the force of inertia acting on it. This process is a mechanism which prevents matter from travelling at or beyond the speed of light. Einstein saw this as a very important feature of his paradigm. Unfortunately, the supposed tests of this relativistic effect have been inconclusive in their analysis. Instead of attempting to measure the mass of rapidly moving particles, these experiments have only confirmed what is known as the *e/m* ratio. This is the famous charge to mass ratio of the electron that was originally measured in 1897 by J.J. Thompson when investigating the nature of the then recently discovered mysterious cathode rays that now lie at the heart of conventional television tubes. His discovery that *e/m* was a constant for cathode rays and was almost 2000 times larger than for atomic hydrogen ions led to his proposal of the first subatomic particle, the electron.

The ratio, *e/m*, became of interest, because for historical and technical reasons it was possible to measure it before either quantity on its own. It took 20 years before the first measurement of the fundamental charge on an electron in 1917 by R.A. Millikan. In 1908, Alfred Bucherer measured *e/m* for electrons moving over a large range of speeds up to 70% of the speed of light. He claimed that *e/m* decreases with velocity in a manner which is in accordance with the theory of SR and this was hailed as a confirmation of the theory.

However, there are at least two other explanations of Bucherer's results. One is that the charge on the particle could decrease with velocity. This possibility was only seriously considered and rejected in 1960, when it was theorized that such a relationship between charge and velocity would make it impossible to guarantee the charge neutrality of atoms made up of slow heavy positive protons and less massive and faster electrons. [11.19]. Another possible explanation has never been publicly considered which is that the electromagnetic forces acting on a charged particle may be affected by the velocity, *v*, of the particle relative to the charged plates or magnets used in the accelerator apparatus. If we entertain the notion that these forces might depend on *v/c* then it becomes impossible to deduce whether it is the force or the mass that is varying with the speed of the particle.

However, the entire motivation for the theory of SR was to attempt to secure the validity of the Maxwell-Lorentz field theory which included the Lorentz law of electromagnetic force on a charged particle which always increases with velocity. Consequently to test SR one must assume the validity of the Lorentz force law and the only explanation of the Bucherer results is that mass varies with speed. This however is pure assumption and it is just as valid to assume that mass remains constant at all velocities and the electromagnetic force on a charged particle decreases as its speed approaches c, the speed of light. This does however require a different electromagnetic force law which is not currently described in any conventional textbook.

The electromagnetic force law proposed by Wilhelm Weber in 1850 would however have supported this unvarying mass interpretation of the Bucherer experiment. It is remarkable that several years before Maxwell began publishing his electromagnetic field ideas, Weber had already introduced the fundamental constant c. As described in chapter 10, he needed this constant in order to unify the physical units in Coulomb's law of electrostatic force and Ampere's law of electrodynamic force. Weber defined c as "the relative velocity for which two electrical masses do not at all interact" [11.20]. Weber and his colleague Rudolph Kohlrausch took five years to measure this constant which they achieved with remarkable accuracy given the experimental equipment of the day. Much to their dismay, they arrived at a figure which was very close to the best measure of the speed of light in 1856. However, they could see no connection between their electromagnetic constant and the speed of light. It was Maxwell, ten years later, who finally connected the action at a distance constant, c, with the concept of electromagnetic radiation which included light. Utilising Weber's definition of c, Newtonian Electrodynamics seems to have anticipated the results of the Bucherer type experiments in which it became evident that it was impossible to accelerate matter faster than c by means of electromagnetic forces.

Since 1908, many similar experiments have been performed that all purport to confirm the varying mass predictions of SR. However all of these claims suffer in the same respect in that they exclude the possibility that electromagnetic forces may well decrease with increasing relative velocity between a charged particle and an external magnet or charged electrode. These type of experiments are therefore inconclusive and certainly provide no evidence for or against SR.

Time Dilation ?

Most of the assurance gained by conventional physicists regarding the theory of special relativity comes from another family of experiments. These are the supposed tests of SR's prediction of time dilation. It is now proved beyond doubt that the speeding up and slowing of clocks and other time measuring devices often occurs as a result of their motion. This effect has a critical role in at least one area of technology which is widely used every day.

With a small hand held receiver, it is now possible to discover one's precise location on earth thanks to GPS (Global Positioning System) satellites developed primarily by the US defence industries. The accuracy of the measurement depends on precise atomic clocks which are in each of the 24 satellites orbiting the earth at an altitude of approximately 20,000 km. In order for the system to work, each satellite broadcasts a unique signature code as well as the precise time of signal transmission. A GPS receiver needs to receive signals from four or more satellites in order to calculate its location in three spatial dimensions as well as time. For a position to be calculated, it is crucially important that all of the satellite clocks are ticking at the same rate. This is achieved by continuously monitoring the positions and clocks of the satellites from six strategically located ground stations. If any of the clocks is found to be keeping incorrect time relative to the ground station, the satellite is advised and readjusts itself. The type of atomic clocks used in this system are generally found to randomly gain or lose a few nanoseconds (billionths of a second) in a day.

There is however a very important difference between the atomic clocks in the satellites and those at the ground station on earth. The clocks that go into space are specifically constructed to lose 38.4 microseconds (millionths of a second) per day compared to the clocks which will remain on earth. Only, then when the clocks are placed into their very specific orbit will they tick at the same rate as the earth based clocks. This is equivalent to taking a normal watch which might perhaps gain or lose one second per week and then engineering it to lose an hour every day. In other words, the clocks that are placed on the satellites run very differently from the earth based ones when sitting next to each other on the lab bench before and after take off. However when the clocks are in orbit, all of the clocks including those on the ground tick at the same rate. This is indisputable evidence that motion affects the performance of clocks in a predictable manner.

This behaviour has been interpreted as a confirmation of Einstein's theory of relativity. However, it cannot simply be a result of SR which predicts that the satellite clock, which is moving faster than the ground clock around the centre of the earth, should tick slower. In fact, as we have just seen, the clock in orbit actually speeds up. The relativistic explanation of the speeding up of the clock comes from Einstein's theory of General Relativity. This aspect is often described as the gravitational redshift and claims that clocks slow down in increasing gravitational fields. Since the satellite clock is further away from the centre of the earth than the terrestrially based clock, it has less gravitational force acting on it and therefore it should tick faster. We see that Einstein's two relativity theories predict two entirely distinct mechanisms that both affect the timekeeping of a clock, one depends on inertial velocity and the other on gravitational force.

Examining the GPS satellite clock more closely, we can say that it is not moving inertially, which would require that it be unaccelerated and under the influence of no external forces. A clock in earth orbit, which is moving in an almost circular trajectory, is always accelerating toward the centre of the earth. In fact, its centripetal acceleration is its only dynamical property. For a GPS satellite in a stable orbit at 20,000 km above the surface of the earth, the acceleration toward the centre of the earth is about 1 m/s^2. The atoms in this clock therefore feel an equal and opposite centrifugal force of inertia equal to their mass multiplied by 1 m/s^2 pushing away from the centre of the earth. Since every part of the satellite feels equal and opposing gravitational and inertial forces, there is no externally produced relative acceleration between any parts of the clock. This physical situation is often called "zero *g*" or "free-fall". It allows astronauts to float freely in their spacecraft and clocks to beat at their natural frequency.

In the general discussion on clocks presented at the end of this chapter, it will be shown that in a clock, not in a zero *g* environment such as on the surface of the earth, part of the internal mechanism that moves relative to the case has an extra acceleration component relative to the case. This occurs because the outside of the clock is being acted on by an external contact force such as the upward reaction applied by the earth's crust. This extra internal acceleration can cause the clock to tick at a frequency that is slightly different from its natural frequency. It is therefore an empirically based hypothesis that the timekeeping of clocks is affected by external forces which affect the relative accelerations of their internal parts. Unlike Einstein's theories, our new hypothesis makes no claims as to the nature of "time", but instead simply describes how the relative ticking frequency between two identical clocks depends on the physical forces to which each is subjected.

Such a theorem explains the discrepancies in timekeeping of identical clocks with a single physical mechanism, namely external force. However, unlike SR, it ascribes no effect on clocks as a consequence of inertial (force free) motion. This is important for as we will see in the next paragraph, none of the supposed "time dilation" experiments have been performed on clocks undergoing anything like inertial motion.

As Dingle had observed, if two clocks are both moving inertially without the influence of any external forces, at most, they can only meet each other once and at this moment, they can compare their readings. They will never meet again and thus one could never by direct measurement actually determine whether one was ticking faster than the other. Einstein's solution to this dilemma was to propose the twin paradox in which one of the twins travels inertially at high velocity and then turns around and returns again moving inertially and finds himself younger than his brother who stayed at home. Needless to say, this experiment has never been performed, but a century long debate has slowly raged about whether the turning around (acceleration) of one of the brothers, presumably by the firing of rockets, made him the one who would be younger upon their reunion. The famous pioneer of quantum mechanics, Wolfgang Pauli [11.21], pointed out that it was not until 1918, that Einstein made brief mention that the acceleration of the traveller during the turn around must be involved, however acceleration or force were not used in his equations of time dilation. The case of the travelling twins reveals the clue that not only are applied force and acceleration critical to determine which clock ticks faster than another, but that acceleration is always required to perform any such experiment.

Textbooks attempt to resolve this dilemma by arguing that if one had a fantastic telescope and could observe a clock moving at a speed near c, then it would appear to be running slow. As a result, modern physics is now completely ambiguous whether the moving clock is actually running slow or whether this is just an outcome of the method of observation. Needless to say, no such telescope yet exists and thus this aspect of SR is also experimentally unproved.

Most of the experiments that purport to demonstrate time dilation involve time keeping instruments moving in circular paths. As with the atomic clocks in the GPS satellites, caesium clocks have also been placed in jet airplanes and flown around the world on commercial flights. A famous experiment was performed in 1971 by Hafele and Keating under the auspices of the United States Naval Observatory (USNO). They published their results in the journal, Science, [11.22] in 1972 and reported that:

"These results provide an unambiguous empirical resolution of the famous clock "paradox" with macroscopic clocks"

This apparent success for Einstein's theories was so widely disseminated that in the same year a leader article in Nature [11.23] claimed that "the agreement between theory and experiment was most satisfactory". The experiment is still famous and described in every modern relativity textbook. However, in 1995, a very enterprising Irish engineer, Alphonsus Kelly, under the US Freedom of Information Act, was able to secure a copy of the originally classified USNO internal report filed by Hafele and Keating in 1971 which included all of their raw data and analysis. [11.24] In this document, Kelly found that as in the Eddington eclipse debacle, much of the data had been left out of the published paper including mention of how they justified huge manual clock corrections, some ten times larger than the measured result, which completely changed the outcome of the experiment. Kelly [11.25] has analyzed these original results closely, but the conclusion drawn by the original scientists is by far the most damning. Hafele wrote in the internal report:

"Most people (myself included) would be reluctant to agree that the time gained by any one of these clocks is indicative of anything ... ...the difference between theory and measurement is disturbing"

Clearly, the Hafele and Keating results should not be used as evidence for or against the theory of SR. What is however much more worrying is that the authors and their supervisors allowed such dishonest science to be published, and said nothing when it became clear that it was being held up by the scientific community as some of the strongest evidence supporting Einstein's relativity theories at that time.

The most striking clock distortion results come from particle accelerators in which very fast sub-atomic particles travel around a storage ring. The best known of this type of experiment was performed at CERN in Switzerland in 1977. Muons are relatively unstable particles which undergo spontaneous decay into electrons and neutrinos. They must have some kind of timekeeping mechanism contained inside them since a group of such muons can be defined as having a repeatable "half-life" which is the statistical time during which we can expect half of them to decay. Such a period can be measured when the muons are at rest relative to the accelerator. However if a collection of these particles is accelerated by large electromagnets in multiple revolutions of a storage ring, the half life is found to increase. This has been interpreted by Einsteinian relativists as the internal time keeping of the particles being slowed down. The scientists, whose results were published in the journal Nature [11.26], made the claim to have slowed down the internal timekeeping of the particles by a factor of 29.3. They achieved this by subjecting very fast muons to a huge centripetal acceleration of $10^{18}g$ in order to keep them in the storage ring. While this result has been cited on many occasions as one of the strongest pieces of evidence supporting the theory of SR, it can easily be seen that the muons are being massively accelerated and therefore cannot remotely represent a clock moving inertially as required by SR. This result would also have been more convincing, if the authors had firstly been able to directly measure the muon velocity. Instead they were forced to calculate the speed using SR theory. Secondly, they were unable to publish an experimental numerical relationship between velocity and lifetime dilation over a range of velocities because their accelerator could only contain muons at a single so called "magic" energy. Therefore it could be a coincidence that their time dilation factor matched the SR prediction using the calculated muon velocity. The analysis presented with these results clearly represents an unjustifiable circular logic in which the validity of SR has to be assumed in order to perform a test on its own predictions. Whether this experiment supports SR or not, it certainly provides yet another strong connection between applied force and the slowing of a clock.

Almost all of the time dilation experiments that have been used to supposedly provide support for SR have involved clocks moving in circular trajectories, whether they be particles in an accelerator or atomic clocks in planes or satellites travelling around the earth. Since, by definition, circular motion involves acceleration and external forces, none of these tests have actually examined the claims of SR which only apply to clocks moving inertially.

There is however one well known supposed test of SR that does not involve circular motion. Like the experiment at CERN, it also involves the use of muons as clocks with a predictable decay half life when at rest relative to the laboratory. This experiment was first performed in the early 1940's by Rossi and Hall [11.27] and then later with more accuracy by two MIT scientists Frisch and Smith [11.28]. In both of these experiments, the goal was to measure the number of particles which have decayed while traversing a known straight flight path at a constant velocity.

In these experiments, they did not detect the actual decay events, but rather measured the number of radioactive particles at two different locations. In the case of the MIT experiment, these sites were the top of Mount Washington in New Hampshire at an altitude of 1930 meters and a lab at sea level in Cambridge Massachusetts. For the purpose of this experiment, these two locations are near enough to each other to be considered at the same place but at different altitudes. The muons are created in the upper atmosphere by the collision of cosmic ray particles from the sun with atomic gas nuclei. These muons fly in all directions including toward the earth surface with speeds greater than $0.99c$. As they descend, some of them decay so that we would expect to find fewer of them at sea level than at the top of the mountain. On top of Mount Washington, they aimed to measure the number of muons with velocities between $0.9950c$ and $0.9954c$. They claimed to achieve this by using a thick piece of steel in front of a thin plastic particle detector. Muons which were too slow would not penetrate the steel and those that were too fast would pass through both the steel and the detector and not be counted. The muons that were detected had therefore been reduced to a negligible velocity by the steel and thus their "low velocity" lifetime in the detector could be measured. In this way, they counted 568 particles per hour with a mean lifetime of 2.2μs at the top of the mountain. They argued that particles travelling near the speed of light would pass through the atmosphere from 1930 meters to sea level in 6.4μs, and based on their measured distribution of lifetimes, they expected only 27 particles per hour to survive all the way to sea level, based on a non-relativistic calculation. However, they report a sea-level measurement of 412 particles per hour. They then claim this as evidence that the particles live longer when travelling at high velocities in accordance with the predictions of SR.

However, this famous experiment which is presented to every physics undergraduate to confirm their belief in SR is tainted with a piece of experimental fudging which is rarely if ever discussed and has been overlooked by a generation of physics teachers. Scandalously, the measurement apparatus at both locations was not in fact identical, for as Frisch and Smith point out that "by the time they (the muons) reached sea level they had been slowed down somewhat by the air". Later in the paper, they estimate that this deceleration amounts to approximately $2\times10^{13}g$. They required this change in velocity as justification for using 40% less steel in front of their detector at sea level compared to the set-up on Mount Washington.

Probably, the strongest promotion of this experiment to students of physics is the textbook entitled *Special Relativity* [11.29] by the author and MIT professor A.P. French. He fills four pages of his book with a detailed description of the Frisch and Smith experiment including still frames from a film made during its operation. In French's presentation of the experiment, he completely ignores the fact that the two detectors, one on the mountain top and the other at sea level are not identical. The fact that such a thorough physicist would omit to describe such a major feature of the experimental set-up is certainly highly suspicious. As a result of the fact that different detectors were used for the two measurements, this test can certainly not be considered to be a controlled laboratory experiment. The argument used by Frisch and Smith to justify their removal of 40% of their steel absorber is entirely based on relativistic kinematics. Unfortunately, it is illogical to conduct a fundamental test of a theory if one is required to assume the validity of the theory in the analysis. Rather, at best such an experiment can only demonstrate that Einstein's theory is mathematically self-consistent which it undoubtedly is. However in no way should these results continue to be promulgated to future generations of students as a valid proof of relativistic time dilation.

There are several other types of experiments that are traditionally accepted as evidence in support of relativistic time dilation. Tests to investigate Einstein's prediction of a relativistic redshift involve the observation of an oscillator such as a light emitting source by a detector in relative motion. If one applies the assumptions of relativity theory in order to interpret the observations, then one can infer that the internal timekeeping of the source appears to have slowed down. However, this analysis requires the use of Einstein's purely hypothetical conjectures regarding the speed of light in order to confirm the effects ascribed to the theory. This is again circular reasoning and cannot be used to prove or disprove SR.

As Dingle would have predicted, in all of the years of trying, nobody has performed an experiment in which an unaccelerated clock has been shown to increase or decrease its elapsed time with respect to another unaccelerated clock. Therefore, empirically, we are drawn to the conclusion that it is applied force and acceleration which affect the internal mechanisms inside clocks whether they be oscillating wheels in mechanical clocks, quartz crystals, electron vibrations in atomic clocks or even faster microscopic beating inside sub-atomic particles. This is the fundamental connection between forces of inertia and the act of timekeeping.

Mach and others quite reasonably objected to Newton's concept of absolute space and time, which historically led to the eager acceptance of Einstein's revolutionary conception of relative time as something that varies for every observer. Both Newton's and Einstein's concepts of time are very difficult to handle philosophically and instead it is proposed here that the concept of time has no meaning at all, since all that we can actually measure and describe are the relative ticking rates of different clocks.

Timekeeping

It can be said that during a series of seasonal events that we traditionally call a year, the moon goes around the earth 12.4 times while the earth spins on its axis 365.24 times. The ratio of these two numbers is also a dimensionless number which is the relative ticking frequency of two large clocks. It is an experimental fact and has nothing to do with units of time such as the second or the year or even the choice of observer. There are many other astronomical events which have been discovered to occur at frequencies which stay roughly constant with respect to each other. As long as these measurements are taken using the apparently fixed distant stars as a background reference, these ratios remain consistent for all reasonably local observers whether they be on the earth or any other planet in the solar system or even in a passing spaceship. Through the centuries, astronomers have used these ratios of periodic events involving objects which are all subject to nominally unchanging forces in order to develop a useful system of units. Once we had an arbitrary unit, such as the year, horologists devised machines which tick a repeatable number of times during one of these years and then called the period of one of these ticks another name. In our age of reliable clocks, the most common unit of time is now the second. Until, the 1950's, clocks were specifically constructed to count seconds in such a way to ensure that there were always 31,557,600 of them in a year. Now the second is defined by a certain number of vibrations of a cesium atom which is even more accurate than the motions of celestial bodies.

During a day the earth revolves once upon its axis with respect to the sun while the second hand on a trusted watch ticks 86,400 times. Can we expect these ratios to always remain constant? The answer is of course that clockmakers have been aware for centuries that mechanical clocks are very sensitive to external conditions. Temperature and humidity variation were major problems for early clock makers. Abrupt shocks which involve significant accelerations can also change the ticking rate with respect to the spinning of the earth. The need to reduce this acceleration sensitivity has been one of the primary drivers of technological innovation in clock design for the last 300 years. By the late 17th century, there was a desperate need for more reliable clocks which were required for navigation at sea and which could work in the harsh environment of constant wave motion, and a wide range of atmospheric conditions. By the middle of the 18th century, John Harrison was awarded a princely £10,000 prize for creating a watch that lost only 5 seconds after 81 days at sea.

A further social change that demanded technological improvement was the introduction of the wristwatch, which was initially a fashion accessory for well to do Victorian ladies who had no waistcoat pockets in which to hold a conventional pocket watch. The quick accelerations of the human wrist compared to the relatively steady torso required further refinements to the delicate springs and balance wheels that make up a mechanical watch. These advances were achieved primarily by decreasing the physical size of the oscillating mechanism and developing more stable and powerful springs which allowed the clock to tick at a faster rate. Harrison's best chronometer oscillated once per second, while a modern mechanical watch may tick up to 10 times per second. There appears to be little use in beating faster because friction and lubrication problems start to become significant. These developments would have been unnecessary unless there was a direct connection between externally applied forces and the ticking frequencies of a clock.

Such problems with wristwatches have now been virtually eliminated by the advent of the quartz crystal watch which almost everyone now uses. They are very cheap to produce and the crystal vibrates fairly reliably 32,768 times per second. This makes them immune to the types of acceleration that humans regularly experience and produce.

In this light, a clock can be viewed not as an instrument for measuring the vague notion of time, but rather a machine designed by nature or man in which one part performs a regular oscillation with respect to the rest of the clock. For the sake of this discussion, we call one part the case and the other the oscillator. By definition, the two must have a non-rigid connection and the only fundamental difference between them is that the case is considered to be the part that directly experiences local contact forces.

For a mechanical wrist watch, the case is clearly the outside of the clock which includes the dial face and is in direct contact with an arm. The oscillator is a finely balanced wheel rotating back and forth inside the case. The frequency of the oscillation is determined by the mass of the wheel and the very accurate reversing force it receives at either end of its oscillation. These parameters in conjunction with the force of inertia caused by the distant universe prescribe the acceleration of the wheel relative to the case at its end stops. This is how the force of inertia acquires a crucial role in the running of a clock.

One way to externally accelerate a mechanical watch is to physically rotate the case around the same axis as the internal balance wheel. It is easy to see that if the case is revolving with the same angular velocity as the balance wheel, it will never reach the next end stop and the watch will have stopped functioning. This is an extreme example of how adding relative acceleration between the case and oscillator can affect the timekeeping of a clock. Fortunately, the rotational motions of the human wrist are not very large, but all mechanical watches with balance wheels are slightly affected by case rotation.

The quartz watch, which has a crystal which vibrates linearly tens of thousands times per second, is much less susceptible to vibration than a mechanical watch. It is also vastly cheaper to make a reliable quartz watch than a mechanical one and as a result it has become the most common time measuring device in the world. These crystals now find themselves in modern communications, navigation and radar systems in which they are subjected to vibrations and accelerations far larger than those produced by a human wrist. For instance a guided missile has navigation equipment aboard that employs quartz clocks, and it is now well known that such a crystal subject to a steady acceleration has a slightly different natural frequency than the same resonator experiencing zero acceleration. [11.30] There is therefore commercial and strategic importance in the precise mathematical relationship between acceleration and frequency for this very important crystal. The amount of frequency shift has been found to be proportional to the magnitude of the acceleration and also dependent upon the direction of the acceleration relative to the axis of vibration of the crystal.

It is therefore reasonable to assume that all clocks are sensitive to acceleration. The higher the frequency of the clock, the less the susceptibility to error. We will not be able to find a relationship between acceleration and frequency shift that is the same for every type of clock since each has a different mechanism of internal oscillation. For instance, the rotation that can stop a mechanical watch will have no effect on a quartz crystal undergoing linear vibration.

We do not know the frequency of the interior oscillation inside a muon, but it is quite reasonable that accelerations of $10^{18}g$, such as found in the CERN experiments, are capable of affecting their ticking frequency and therefore their average lifetime. At first, one might think that like a satellite clock in earth orbit, a muon in an accelerator ring is in a zero g environment. However, in a satellite clock, every piece of matter experiences the same centripetal acceleration due to the equal and opposite forces of gravity and inertia. Contrastingly, the very small constituent particles that make up a muon have different electric charges (positive, negative or neutral) and thus react to the external accelerator magnets differently. This must lead to differing internal accelerations inside the muon than those that occur when the particle is not subject to external electromagnetic forces. Therefore we can understand qualitatively how the ticking rate of a muon can be affected by circular motion in a storage ring.

Similarly, in an atomic clock on the earth, all of the matter inside the clock feels the downward force of gravity, but only part of the clock that is in direct contact with the earth feels the upward reaction force caused by the surface of the earth. Therefore, there is a small but real extra acceleration between case and oscillator that does not exist in an identical clock in orbit in a zero *g* environment. This is probably the real reason that a GPS satellite clock runs faster in orbit than an identical clock on earth.

Unlike their velocity, self contained systems can directly detect their own acceleration without reference to another nearby body. This can be achieved with accelerometers which take advantage of the forces of inertia which are caused by direct interaction with the distant universe as proscribed by Mach's principle. For instance a person can act as an accelerometer. If he is in a closed box without windows, the acceleration of the box can be determined by measuring the change in the force between him and the floor with a set of bathroom weighing scales. We now see that clocks can be viewed as another type of accelerometer. This book has demonstrated that the forces of inertia can produce very real effects such as pulling rubber off a tyre on a car in a high speed turn, or compressing the springs of a weighing scale. Now we know that the force of inertia can also cause a shift in the ticking frequency of a clock.

Contrary to the conventional presentation in textbooks on SR, it is now apparent that the frequency distortion of clock mechanisms can be attributed to acceleration caused by the application of external forces. This experimentally based explanation of some of the very real and observer independent clock effects, that until now have been described as "time dilation", frees us from the unverifiable philosophical confusion imposed by SR in which physical effects are assumed to depend only on inertial velocity which has a different value for every possible observer. It now seems that both the timekeeping of clocks as well as the strength of the force of inertia depend on acceleration which as we have seen throughout this book can only be accurately described in relation to the background of fixed stars, implied by Mach's principle.

Therefore in Machian philosophy, timekeeping is analogous to acceleration itself and only has meaning if the entire universe is involved. Mach's principle describes a background of stars and galaxies that are so far away from us that on the timescales of any measurement we are ever likely to make the stars remain fixed in space. We need this apparently unmoving distribution in order to meaningfully compare our measurements of acceleration. Similarly, we can also imagine a background of distant clocks, ranging from vibrating sub-atomic particles to galaxies orbiting around each other. This ensemble can be considered from the point of view of our human sized experiments to be working with constant relative ticking frequencies. We need such a group in order to compare the changes in ticking frequency in clocks near us when subjected to external forces. Thus a virtually steady Machian background composed of real objects is essential for the understanding of clocks.

Einstein went to great trouble in order to save the Maxwellian theory of electrodynamics and left us with a philosophy based purely on relative motion between nearby objects. He hoped that the behaviour of clocks would eventually prove his theory to be physically real. We have now seen that no controlled experiment has yet confirmed Einstein's relativistic theories. Instead, all we have is clear evidence that acceleration can cause clocks to change their frequency.

By maintaining and comparing large numbers of clocks, mankind has been able to create the comfortable feeling of measuring the passage of time, a process usually called timekeeping. When most clocks remain in synchrony, but one is found to change its relative ticking frequency to the rest, then this can usually be related to an acceleration of the particular clock caused by an external force. (Sometimes however it is simply an unpredictable failure of an internal part.) This does not require time be a fundamental property of the universe. Instead, the act of timekeeping is a human activity which requires counting as well as an understanding of force and acceleration. Mach's principle makes clear that in order to preserve the principle of momentum conservation, the relationship between applied force and acceleration is controlled by forces of inertia caused by the self-interaction of the entire universe. Therefore timekeeping and the study of clocks are fundamental aspects of our understanding of the force of inertia.

Chapter 11 References

[11.1]   C. L. Poor, "The deflection of light as observed at total solar eclipses," *J.Opt.Soc.Amer*, vol. 20, p. 173-211, 1930.

[11.2]   A. P. French, (Ed.) *Einstein: a centenary volume*. Cambridge, MA: Harvard University Press, 1979.

[11.3]   R. Nadeau, M. Kafatos, *The Non-local universe*. Oxford: Oxford University Press, 1999.

[11.4]   G. Burniston-Brown, "What is wrong with relativity?," *Bulletin of the Institute of Physics and Physical Society*, vol. 18, p. 71-77, 1967.

[11.5]   L. Essen, "Relativity and time signals," *Wireless World*, vol. 83(October), p. 44-45, 1978.

[11.6]   H. Samuel, H. Dingle, *A Threefold Cord, Philosophy, Science and Religion*. London: Allen & Unwin, 1961.

[11.7]   A. Einstein, "Zur Elektrodynamik bewegter Körper," *Ann.Phys*, vol. 17, p. 891-921, 1905.

[11.8]   A. Einstein, H. A. Lorentz, H. Minkowski, et al., *The Principle of Relativity*. London: Methuen, 1923.

[11.9]   H. Dingle, *Science at the Crossroads*. London: Martin Brian & O'Keefe, 1972.

[11.10] P. Graneau, N. Graneau, *Newtonian electrodynamics*. Singapore: World Scientific, 1996.

[11.11] A. A. Michelson, E. W. Morley, "On the Relative Motion of the Earth and the Luminiferous Ether," *American Journal of Science*, vol. 36(203), p. 333-345, 1887.

[11.12] D. C. Miller, "The Ether-drift experiment and the determination of the absolute motion of the earth," *Review of Modern Physics*, vol. 5, p. 203-242, 1933.

[11.13] A. A. Michelson, H. A. Lorentz, D. C. Miller, et al., "Conference on the Michelson-Morley Experiment," *The Astrophysical Journal*, vol. 68(5), p. 341-402, 1928.

[11.14] R. W. Clark, *Einstein, the life and times*. New York: World Publishing, 1971.

[11.15] R. S. Shankland, S. W. McCuskey, F. C. Leone, et al., "New analysis of the interferometer observations of Dayton C. Miller," *Rev.Mod.Phys*, vol. 27(2), p. 167-178, 1955.

[11.16] T. S. Jaseja, A. Javan, J. Murray, et al., "Test of Special Relativity or of the isotropy of space by use of infrared masers," *Physical Review*, vol. 133(5A), p. 1221-1225, 1964.

[11.17] M. Allais, "The origin of the regularities displayed in the interferometric observations of Dayton C. Miller 1925-1926: Temperature effects or spatial anisotropy?," *C.R. Acad.Sci.Paris*, vol. t.1, serie IV, p. 1205-1210, 2000.

[11.18] T. E. Phipps, *Heretical verities: Mathematical themes in physical description*. Urbana, IL: Classic Non-Fiction Library, 1987.

[11.19] W. G. V. Rosser, *Relativity and high energy physics*. London: Wykeham Publications, 1969.

[11.20]   O. Darrigol, *Electrodynamics from Ampere to Einstein*. Oxford: Oxford University Press, 2000.

[11.21]   W. Pauli, *Theory of Relativity*. New York: Dover, 1981.

[11.22]   J. C. Hafele, R. E. Keating, "Around the world atomic clocks (predicted & observed) relativistic time gains," *Science*, vol. 177, p. 166-170, 1972.

[11.23]   Leader, *Nature*, vol. 238, p. 2425, 1972.

[11.24]   J. C. Hafele, R. E. Keating, "Proc. 3rd Dept. Def. PTTI Meeting," USNO (United States Naval Observatory), 261-288 (1971).

[11.25]   A. G. Kelly, "Hafele & Keating : Did they prove anything ?," *Physics Essays*, vol. 13(4), p. 616-621, 2000.

[11.26]   J. Bailey, K. Borer, F. Combley, et al., "Measurements of relativistic time dilation for positive and negative muons in a circular orbit," *Nature*, vol. 268, p. 301-305, 1977.

[11.27]   B. Rossi, D. B. Hall, "Variation of the rate of decay of mesotrons with momentum," *Phys.Rev.*, vol. 59(3), p. 223-228, 1941.

[11.28]   D. H. Frisch, J. H. Smith, "Measurement of the relativistic time dilation using mu-mesons," *Am.J.Phys*, vol. 31, p. 342-355, 1963.

[11.29]   A. P. French, *Special Relativity*. London: Chapman & Hall, 1968.

[11.30]   R. L. Filler, "The acceleration sensitivity of quartz crystal oscillators : a review," *IEEE Trans. Ultrasonics, Ferroelectrics and Frequency Control*, vol. 35(3), p. 297-305, 1988.

Chapter 12

Machian Inertia and the Isotropic Universe

A New Force Law

Throughout the previous chapters of this book, we have described how the force of inertia is very likely to be caused by instantaneous interactions between all pieces of matter throughout the universe. However, this concept, generally known as Mach's principle, cannot be directly proven. Like all cosmological models, it suffers from the fact that we have no means to alter the distribution of material throughout the universe in order to truly test our hypothesis in a controlled manner. However, this failing applies to all cosmological models and is therefore no reason not to speculate.

In this chapter we are proposing a novel Newtonian force law which is consistent with all experience and describes the force of inertia in terms of a universal constant, which represents the distribution of mass throughout the universe. The full mathematical justification for this entirely Newtonian force law was recently published in the journal, "General Relativity and Gravitation" [12.1]. It develops a model of inertia and gravitation which is entirely based on instantaneous action at a distance. The mathematics is not appropriate for this book and therefore only a general discussion of the argument and conclusions will be presented here.

Newton's laws of motion imply that the force of inertia counteracts acceleration. In conjunction with Mach's principle, a law describing the force of inertia must predict that a particle which accelerates in any arbitrary direction in the midst of an isotropic mass distribution, will experience a repulsion from half the distribution of matter which is in front of it and an attraction from the other half which is behind it. Isotropic means that the matter distribution looks the same in all directions. These repulsions and attractions can combine to create a finite force of resistance to acceleration.

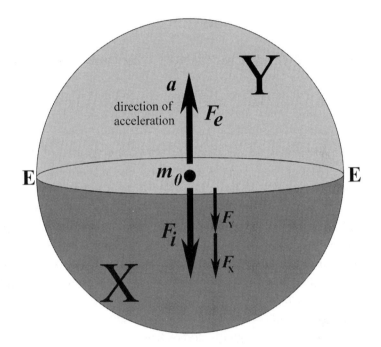

Figure 12.1 : Depiction of interactions between a particle experiencing an external force and all of the objects in the universe, creating a force of inertia

Consider the diagram in figure 12.1 in which a particle of mass, m_0, in the laboratory is being acted on by an upward external force, F_e. If the particle is free to move, it will accelerate with respect to the fixed stars (Machian inertial system) in the direction of the applied force. If the inertial force, F_i, is proportional to the magnitude of the acceleration, a, and acts in the opposite direction, then it will increase from zero as soon as the particle begins to accelerate. The inertial force increases as the acceleration increases, ensuring that the force of inertia is always equal and opposite to the applied external force. This robust balancing act was described by D'Alambert as dynamic equilibrium. If a particle were to spontaneously accelerate without the action of an external force, then the inertial force would arise and immediately negate this extra acceleration. This stability is the mechanism behind Newton's first law which states that an object cannot accelerate with respect to the distant universe unless acted upon by another body.

A Newtonian force of inertia should have the same form as Newton's law of gravitation as well as the other famous Newtonian law, Coulomb's law of electrostatic force. Less famous, but equally accurate are two other Newtonian force laws, namely, Michell's law of the force between magnetic poles and Ampère's electrodynamic force between current elements. All of these laws have several features in common. They represent attractions or repulsions between separated objects. The strength of these forces is proportional to the product of a specific property of each object. In the case of Newtonian gravity, this is mass and for Coulomb's law it is charge. All of these force laws also contain an inverse square dependence on the distance of separation. This means that their strength decreases rapidly as they get further apart. Each one also contains a mysterious dimensional constant which makes the force agree with experiment. These constants have been very carefully measured but it is still unclear what determines their magnitude. Perhaps this chapter can shed some light on where the value of Newton's gravitational constant, G, actually comes from.

A New Force Law

The law of inertia that we are proposing is another member of this small club of Newtonian force laws. It is most closely related to Newton's law of gravitation in that it is proportional to the product of the masses of the two interacting bodies. Its unique feature is that it depends on the relative acceleration of the two objects, a. If r is their distance apart then the inertial force between two masses, m_0 and m_x can be expressed as

$$\Delta F_i = -\frac{1}{\pi^2 B} a \frac{m_0 m_x}{r^2} \quad .$$

B is a dimensional constant whose meaning will be addressed later in this chapter. If this force is negative it represents an attraction between the objects and conversely a positive force describes repulsion. While mathematical equations have been deliberately left out of this book, it is hoped that the reader will share our enthusiasm that the first new Newtonian force law proposed for over 180 years deserves to be printed.

Let us look closely at what this law predicts for the forces acting on any general particle such as m_0 in figure 12.1. If a force, F_e, is applied to it, then the direction of this force defines a geometrical plane which is perpendicular to the force and also passes through m_0, which is depicted by the plane **EE** in figure 12.1. This allows us to divide all of the matter in the universe into two groups divided by **EE**. In figure 12.1, all of the objects below the plane (**X**) can be described as behind m_0 and those above the plane (**Y**) as in front of it. As a result of the force, F_e, the particle, m_0, accelerates upward, thereby increasing its distance from all of the **X** objects and decreasing its distance to the **Y** objects. This means that m_0 has a positive acceleration relative to everything in the **X** hemisphere which results in an attractive force of inertia between it and every object behind it, F_X. All of these individual forces due to **X** material clearly pull back on the particle to resist the applied force. Conversely, the negative relative acceleration between m_0 and all the **Y** objects in the upper hemisphere result in repulsive inertia forces which push backwards on the particle and combine to counteract the applied force, F_Y. In this manner, all of the objects in the universe cause an inertial force on m_0 which resists its acceleration, F_i. If the matter in the universe is isotropically distributed then this symmetry compels the sum of the individual forces of inertia acting on m_0 to be precisely counter-aligned with the applied force.

Our local pocket of the universe is comprised of relatively few objects distributed quite non-uniformly throughout the sky. This anisotropic distribution includes our solar system and many of our relatively nearby galaxies. However, astronomers now, and mathematicians and philosophers for many years before, have declared that at some very large distance, the distribution of matter becomes random enough to declare that the universe is essentially isotropic. This means that if one looks out far enough, the universe appears the same in all directions. As a result of the vast and possibly infinite scale of the universe, we can assume that virtually all of the mass of the universe is tied up in this isotropic distribution. It follows that as far as Mach's principle is concerned, the material in our local anisotropic region of the universe is insignificant.

Everyday experience tells us that the force of inertia has the same magnitude as the applied force and always acts in exact opposition, irrespective of the direction with respect to the fixed stars. The mathematics presented in [12.1] demonstrates that in order to produce this predictable behaviour, our proposed mechanism relies on the reality of a universal isotropic distribution of matter. Given this symmetrical model of the cosmos, our simple Newtonian force law yields a physical mechanism that explains the force of inertia. Most importantly, it shows that inertia need not be written out of physics for lack of an identifiable cause, as is presently thought to be the case.

Our model does of course assume that the interactions between all objects are mutual and instantaneous, which violates one of the fundamental principles of Einsteinian relativity. However as described in Chapter 1, scientists from Laplace to contemporary astronomers like Tom Van Flandern have revealed experimentally that the speed of propagation of a central Newtonian gravitational attraction is at least $2 \times 10^{10} \, c$, where c is the speed of light [12.2, 3]. Such a high velocity is experimentally indistinguishable from an instantaneous interaction. Consequently, we feel justified to assume that instantaneous mutual interactions could be the cause of the force of inertia.

The constant, B, in the inertial force law given above is a quantity that our derivation shows is related to the total mass of the universe. It describes a modified summation of the mass of every single object in existence. With reference to the particle on which we are calculating the force of inertia, every other object in the universe can be defined by its mass as well as its distance of separation. In this way, each body in the universe can be given a value, B_i (mass divided by the square of distance (kg/m^2)), with respect to the position of a specific object. B is simply the sum of the B_i values for every other object in the universe.

The most serious concern with our proposed inertial force law is the possibility that if the universe is infinite in extent and contains an infinite amount of matter then B may have an infinite value. This would cause the value of the inertial force between an accelerating test object and any other body in the universe to be zero. However since the total number of objects in the universe is infinite, then the total force on the test particle would be undefined, rendering our proposed force law useless. However, it is now clear from recent unexpected astronomical observations that in fact even in an infinite universe, B can have a finite value and as a result our law becomes numerically meaningful.

The Fractal Universe

In the last 30 years, astronomers have discovered that contrary to the assumptions of the previous 300 years, the universe is not homogeneous, but appears to have a hierarchical structure which is best described as a fractal. Shapes and distributions that can be characterized by the mathematics of fractals have been described in a plethora of recent popular scientific literature. Posters of the now famous fractal Mandelbrot curves have become very fashionable and are worthy of artistic merit. Fractal mathematics has been used by scientists in areas as diverse as weather modelling and understanding the formation of coastlines and patterns in plants. The characteristic that defines a fractal is that it involves a pattern which is the same at any scale of magnification or reduction. An example of a fractal distribution that can help us understand the structure of the universe is shown in figure 12.2. The obvious recurring pattern is the distribution displayed on the 5 surface of a normal die. In figure 12.2, the pattern can be seen on 3 different scales, and it can be conjectured that it could continue onto larger scales. As well, we can imagine that each of the dots could be magnified to reveal this same distribution at smaller scales as well.

The most relevant feature of figure 12.2 to our fractal model of the universe is that as one moves away from the centre, the picture becomes more and more dominated by empty space. Therefore, one could say that the density of dots decreases with distance from the centre. Since the 1970's, astronomers have been discovering that not only does the density of the universe decrease with distance away from us in all directions [12.4-6], but also that matter tends to form into clumps such as galaxies and clusters of galaxies and even super-clusters of galaxies. Fortunately, this reduction in density appears to occur roughly at the same rate in all directions. Therefore, we now know that that the universe is not homogeneous even though it still remains isotropic. *B* is therefore a function of not only the density of matter in the universe, but of how quickly this density decreases from our vantage point.

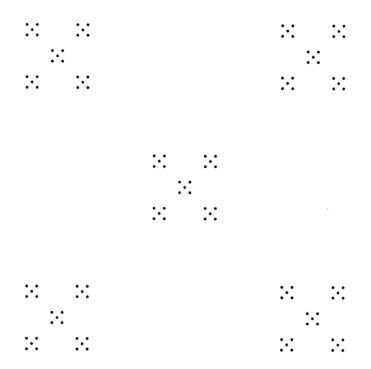

Figure 12.2 : An example of a fractal distribution

Since the time of Galileo, who overturned the earth centred Ptolemaic vision of the cosmos, we have been assiduously conscious that we are not occupying a privileged position in the universe. Consequently, until recently, it has been assumed that the universe had to be a fairly homogeneous distribution of matter with a constant density. Newton was aware that he was caught between two awkward universe scenarios, namely a) that it consisted of a finite amount of matter in an infinite space or b) the apparently atheistic viewpoint that the universe was infinite in extent. The first, (a) would imply that the universe should have collapsed as a consequence of his own law of gravitation and model (b) was unsatisfactory from both a theological standpoint as well the fact that it made the force of gravity acting on a particle due to the entire universe an undefined value. Newton concluded that the delicate stability that we observe must therefore be evidence of the continuous guiding influence of the hand of god. The debate regarding the validity of these two models has rumbled on to this day.

Einstein recognized that Newtonian gravitational theory fell to pieces in a homogeneous universe and attempted to resolve this crisis with his own model. He had a correspondence with an Austrian cosmologist, Franz Selety who was the first to propose an inhomogeneous fractal matter distribution which had no centre. Einstein apparently refused to accept Selety's model because he felt it was not consistent with Mach's principle [12.6]. Even though he clearly understood the necessity of an isotropic distribution of matter prescribed by Mach's principle, Einstein failed to perceive that with Selety's model it was possible to have a centreless distribution which had decreasing density isotropically from any vantage point. Einstein's proposed model therefore retained the classical concept of homogeneous matter distribution but invoked a geometry of curved space in which the mass contained in the universe was finite in order to ensure finite gravity forces but yet the universe was unbounded. His curved space-time was specifically required to ensure that all of the space could be filled with the finite homogeneous matter distribution and yet had no outer edge.

It is not surprising that Einstein was not convinced by Selety's model in the early years of the 20th century as it was still based on conceptual hand waving. The mathematics of fractal geometry lay dormant for many years until developed primarily by the man who actually coined the word "fractal", Benoit Mandelbrot. Born in Poland, and raised in France, he developed an interest in the mathematics of "roughness". While studying the clustering of errors in telephone channels for IBM, he made a connection with theories of inhomogeneous galaxy distributions which had been proposed to explain issues such as why the sky is dark at night (Olber's paradox). It eventually led to the exposition of his "Conditional Cosmological Principle" which was now based on solid mathematical foundations. This principle states that in an isotropic fractal matter distribution with decreasing density, all observers see similar cosmic landscapes around them on the condition that they make their observations from a structural element of the distribution, namely from inside a galaxy.

The convergence of the further development of fractal mathematics and the observations which continue to paint a fractal picture of the universe which extends to further and further distances every year, make a convincing case that the universe is not homogeneous. One can only conjecture that if Einstein had seen Mandelbrot's mathematics, he may never have gone to the trouble of developing his theory of curved space-time, namely General Relativity. If he had then seen the observational evidence for the fractal distribution of cosmic matter, he would have known he was on the wrong track for his theory is fundamentally dependent on the homogeneity of the universe.

The charm of Newtonian gravity is that it can be applied to any matter distribution using simple Euclidean geometry. Newton himself, was aware that there were problems with his universal model, but presumably would be delighted to discover that the universe has an inhomogeneous fractal distribution and that his law of gravity could now produce sensible forces even if the cosmos turned out to be infinite.

Returning to the force of inertia, a fractal distribution of galaxies is still not sufficient to guarantee that B has a finite value. This critical feature depends on the rate at which the density decreases with distance. This rate can be quantified by a number called the fractal (or fractional) dimension, D, which can take on any value between 0 and 3. $D = 0$ represents an single particle universe and $D = 3$ describes a homogeneous distribution. In order for our proposed dimensional constant to be finite, the universal fractal dimension must be less than 2. All of the observational evidence so far indicates that galaxies are distributed with a dimension around the value of 2, but the precision is not good enough to clearly determine whether it is necessarily less than 2. However it is at least conceivable that the actual distribution of universal mass may be consistent with a finite B, which is required to make our proposed force law viable.

There is a strong tradition of ultra-speculative conjecture in cosmology and on the issue of the cause of inertia. In this vein, the English cosmologist Fred Hoyle [12.7] as well as Benoit Mandlebrot [12.8] have both claimed a connection between the inverse square nature of the law of gravitation and the fractal $(D \sim 2)$ structure of the universe. They have gone so far as to suggest that it may be the force of gravity that actually creates the $(D \sim 2)$ structure. It is equally probable that our proposed law of inertia is also closely connected to the observed $(D \sim 2)$ structure. Random galactic motions may be constantly trying to achieve a homogeneous distribution $(D = 3)$, but the universe cannot get beyond $D = 2$ because of forces of inertia.

From a more mathematical standpoint, our proposed law of inertia is so closely related to Newton's law of gravitation, that we can make the claim that if every object was accelerating away from every other with an acceleration of $(\pi^2 BG)$, then our proposed force of inertia is identical and may possibly be the force of gravity. Here, B is the constant in our proposed force law and G $(6.67 \times 10^{-11} \text{ m}^3 \text{ kg s}^{-2})$ is Newton's gravitational constant. With ever increasing cosmological observations, it will eventually be within our powers to estimate B, and then our local laboratory determination of G may come to be the measurement of a universal expansion acceleration. This acceleration may be the mechanism by which the universe avoids becoming homogeneous and retains its hierarchical structure. Recently, several research groups [12.6] have published observations which indicate that very distant supernovae are fainter than expected and this has been construed as evidence that matter in the universe is indeed accelerating away from itself. Therefore it is possible that our proposed inertia force law may well be simply the result of gravity in an accelerating universe.

There is currently great debate concerning the mechanism for this universal acceleration with some people describing it as a result of "dark energy" or "quintessence". Such concepts are required if one is attempting to stay consistent with field theories with finite interaction distances controlled by the speed of light such as exist in General Relativity. However, in a Newtonian instantaneous action at a distance model such as the one described in this chapter in which all cosmic matter can interact with itself instantaneously, we have no way of attempting to apply our limited earth-based understanding to the universe's ultimate behaviour. It seems quite possible that all matter accelerating apart may simply be the natural way in which an infinite amount of matter exists in an infinite space. However as previously declared this is speculation that may go further than is warranted.

A Return to the Newtonian Paradigm

The important and more realistic feature of our proposed Newtonian force of inertia is that it complements Newton's universal law of gravitation and thereby completes Newton's theory of instantaneous action at a distance mechanics in a manner which answers the cosmological doubts of both Mach, who rejected absolute space, and Einstein, who realised that Newtonian gravity was inconsistent with a homogeneous universe. In the early years of the 20th century these concerns led to the development of the theory of General Relativity and the denouncement of Newtonian physics. We now see that this revolution was unnecessary. Recent knowledge of the hierarchical structure of the universe and the consequent finite nature of our proposed inertial force law opens the door for a return to a simpler cosmological model based on Newtonian forces between pieces of matter, acting in a Euclidean geometry. It is important to remember that Newtonian forces and Euclidean geometry have never been found to be in error in any laboratory controlled experiment and are still used with complete accuracy to predict the motion of all man-made objects in our solar system. A famous apparent failure of Newtonian mechanics is the anomalous precession of the perihelion of the planet Mercury. However, this is an example of an uncontrolled experiment in which variables such as solar oblateness and mass distribution cannot be independently manipulated and thus it lacks the rigor with which Newtonian theory has been favourably evaluated.

As a result, the authors of this book suggest that there is strong evidence that the force of inertia is caused by an infinite number of elemental interactions between an object which experiences a force and all of the other pieces of matter in the universe in accordance with Mach's principle. These interactions are manifested as attractions or repulsions and always oppose the applied force. Employing the now well confirmed isotropic fractal distribution of matter consistent with ($D < 2$), the finite magnitude of the force of inertia occurs despite the infinite number of non-cancelling instantaneous interactions. The mass dependent force of inertia is therefore responsible for controlling the magnitude of the accelerations that are caused by applied forces and is the mechanism that lies behind Newton's 2nd law of motion.

We naturally feel that our lives are most controlled by the things nearest to us. Our clothes keep us warm and cars move us around quickly. At further distances, the moon causes the tides and we can even identify the sun as the cause of our comfortable climate. However the planets in our solar system have no detectable influence on physical events on the earth. It is fairly easy to see that as a result, philosophers and now astronomers look at the vast universe simply as a beautiful creation, which is fascinating to observe and can reveal clues as to our history and potentially our fate, but it is generally assumed that we are not directly affected by it. The evidence presented in this book, supporting Mach's principle as the cause of inertia, paints a revolutionary new philosophical outlook. The universe is not simply an aquarium into which we peer, but directly touches us and instantaneously controls the acceleration between every pair of objects. It is hoped that the mental picture of a connected universe will facilitate new cosmological understanding. More romantically, however, this model will also induce a slightly different emotion in the minds of stargazers who should now feel more of a participant rather than just an observer when watching a clear night sky.

Chapter 12 References

[12.1]   P. Graneau, N. Graneau, "Machian Inertia and the Isotropic Universe,"
          General Relativity & Gravitation, vol. 35(5), p. 751-770, 2003.
[12.2]   P. S. LaPlace, *Mècanique Cèleste*, vol. IV. Boston: Little Brown, 1839.
[12.3]   T. van Flandern, "The speed of gravity - what the experiments say,"
          Phys.Lett.A, vol. 250, p. 1-11, 1998.
[12.4]   G. de Vaucouleurs, "The case for a hierarchical cosmology," *Science*, vol. 167,
          p. 1203-1212, 1970.
[12.5]   F. S. Labini, M. Montuori, L. Pietronero, "Scale invariance of galaxy
          clustering," *Phys.Rep*, vol. 293, p. 61-226, 1998.
[12.6]   Y. Baryshev, P. Teerikorpi, *Discovery of Cosmic Fractals*. New Jersey: World
          Scientific, 2002.
[12.7]   F. Hoyle, "On the fragmentation of gas clouds into galaxies and stars,"
          Astrophys.J, vol. 118, p. 513-528, 1953.
[12.8]   B. Mandelbrot, *The Fractal Geometry of Nature*, 3rd ed. New York, N.Y.:
          W.H. Freeman, 1983.

Index